Red Army Legacies

Col. Richard N. Armstrong

RED ARMY LEGACIES

Essays on Forces, Capabilities, and Personalities

Col. Richard N. Armstrong

Schiffer Military History
Atglen, PA

Dedication
To the U.S. Army Military Intelligence Corps - soldiers who live in one army and must know others.

Book Design by Robert Biondi.

Library of Congress Catalog Number: 95-67622.

Printed in the United States of America.
ISBN: 0-88740-805-2

We are interested in hearing from authors with book ideas on related topics.

Published by Schiffer Publishing Ltd.
77 Lower Valley Road
Atglen, PA 19310
Please write for a free catalog.
This book may be purchased from the publisher.
Please include $2.95 postage.
Try your bookstore first.

Contents

Part I: Forces

Part II: Capabilities

Part III: Personalities

ACKNOWLEDGEMENTS

These essays are as originally published or have been revised for this volume as indicated below.

Chapter Two. "The Bukrin Drop: Limits to Creativity," from *The Journal of Military History* (formerly *Military Affairs*), Vol 50, No 3 (July 1986) reprinted by permission has updated material.

Chapter Three. Originally titled, "The Mobile Group Experience," from *ARMOR*, Sep-Oct 1987, reprinted by permission.

Chapter Four. "Mobile Groups: Prologue to OMG," from *Parameters*, Vol. XVI, No. 2 (Summer 1986), reprinted by permission.

Chapter Five. "Hunting Tongues," from *The Journal of Soviet Military Studies*, Vol. 2, No. 4 (December 1989), reprinted by permission.

Chapter Six. "Battlefield Agility: The Soviet Legacy," from *The Journal of Soviet Military Studies*, Vol. 1, No. 4 (December 1988), reprinted by permission.

Chapter Seven. Originally titled, "The Enemy Will Tell Us When," from *Military Intelligence*, June 1981, reprinted by permission. Has been rewritten for more historical content.

Chapter Eight. "Repelling Counterattacks and Counterstrikes," from *Military Review*, July 1989, reprinted by permission has been recasted with more historical examples.

Chapter Nine. "Nachalnik Razvedki: The Red Two," from *Military Intelligence*, January-March 1989, reprinted by permission has been updated.

Chapters One, Ten, and Eleven are previously unpublished studies. All these essays had the advantage of professional peer support. In particular, Ralph Peters encouraged the idea for this volume.

PREFACE

It is a curious thing, in military history; we believe victors often write and shape historical conclusions. But despite battlefield triumphs, the Red Army has suffered defeats at the hands of western historians. Fighting in the Second World War on the Eastern Front offers a vision of inept Red Army hordes overwhelming a superior German Army. This portrayal is firmly etched in our minds primarily because that history has been written by the vanquished. Our inherited German interpretation of fighting on the Eastern Front denied Red Army flexibility to conduct other than mindless echeloned attacks. Use of forward detachments and mobile groups which were forerunners for current armored warfare operations were lost on western analysts. Western armies found it easy to fashion their fighting doctrine against an unthinking, battle-drilled, unimaginative foe.

Upon closer scrutiny of official German military accounts one sensed a healthy respect for the Red Army in daily war journals. These accounts by lower level officers who had no stake in alibiing the loss because of winter snow, spring mud, or Hitler's edicts revealed deeper concerns and acknowledgements of Soviet military prowess.

As a military intelligence officer preparing for the possibility of confronting the Soviet Army, I pursued the clues and sought to reconcile inconsistencies in German accounts. In the study of Soviet military literature, I became aware of how fundamental their military experience in World War II was in shaping their tactical and operational concepts against which I helped prepare U.S. Army units in training and contingency planning.

To be in one army and study another offered an enriching base for understanding military operations. My contributions in war councils were from comparative analysis, suggesting different alternatives to the situation. And, I attempted to share these historical insights in a wider forum through professional writings.

In addition to being a military intelligence officer, I held the additional specialty, historian. I find the skills of an intelligence officer and historian to be similar. Both must establish sources and bona fides of information. Both demand holding a wide range of facts and general knowledge. They must synthesize large amounts of information into an assessment, articulated succinctly and clearly. With an appreciation for past events through historical research comes a sense of perspective. A trained perspective incorporates the ability to enter situations, or problems, with objectivity, allowing one to discriminate between relevant and irrelevant issues. Such a skill is necessary for successful intelligence staff work and enhances a personal ability for critical thought.

My most enduring professional application of military history began in reading Sun Tzu long ago. His 'signs of significance' for indicating enemy's intentions based on human nature held my imagination. Later, I read David Kahn's *Hitler's Spies*, and found a similar listing of 'signs,' or indicators, compiled by the German Army on the Eastern Front during World War II. The value of these human indicators outside unit tactical activities, increased my awareness as an intelligence officer. With a perspective on intelligence reports broadened by historical precedents, I approached my tasks with more depth which grew through the years. A contemporary application in the study of indicators came in the Yom Kippur War of 1973. An Israeli reconnaissance patrol discovered a 'seam' between the Egyptian Second and Third Armies. This crucial piece of intelligence allowed Israeli Defense Forces to conduct a daring cross-Suez Canal attack, carrying Israeli forces into North Africa. This action brought the Egyptian Army to capitulation. It became a personal pursuit to find out how the reconnaissance patrol knew they had a seam between the two armies. Years later, I met an Israeli officer who commanded a battalion in the Suez crossing and asked him about the report. His simple, but astonishing, reply galvanized my historical perspective. The patrol did not know the seam represented a boundary between two Egyptian armies. But, they knew they had a boundary between two different units. They had observed a marked difference in discipline between the units from the manner by which they tied down equipment and canvas on their tactical vehicles. An indicator worthy of Sun Tzu's ancient 'signs of significance' reminded me of the value of historical continuities in immediate actions on the battlefield.

The collection of essays for this book are those that dealt most with historical examples. Some essays I have left unaltered; others I have updated or rewritten for more historical content. And, I have included three essays which never reached publication, although parts were extracted for use in other published pieces.

For a quarter of a century, it was my task to watch and study the Soviet Army for a possible conflict. As a basis for self-study, I tried to understand the Soviet Army's use of its combat experience. One cannot change the past. However, one's view of it, what perspective one chooses to communicate, is dynamic and insightful. My labor was to know how the Soviet Army used its Red Army Legacy.

Richard N. Armstrong
1 December 1994

PART I:
FORCES

CHAPTER 1

Guards of Destruction

As of July 1941, a month after the start of the German invasion, the Red Army had failed to blunt the advance. The battle for Smolensk, nearly 700 kilometers inside the Soviet Union, the German Army's first major objective on their way to Moscow, was about to begin. Marshal Semen Konstantinovich Timoshenko, commander of the Red Army Western Front, reported to Stalin that the Front had insufficient forces and no tanks to halt advancing German panzer units. Fighting to stabilize a retreating front, Red Army forces needed time to slow the German advance. In the face of its seemingly impossible situation, the Western Front headquarters received a cryptic telephone call from the Supreme High Command: "It is planned to use 'ar-esses' against the Nazis, and we have to test them in combat. You will have one experimental unit to test them and report your judgment."[1] Staff officers at the Western Front command post wondered what these "ar-esses" were, what they were supposed to do, and how to use them.

Marshal Timoshenko, on the eve of war as the People's Defense Commissar, knew the secret. On the morning of 17 June, he, along with the People's Commissar for Armament, D.F. Ustinov, and Chief of the General Staff, General G.K. Zhukov, witnessed a demonstration of a new model Red Army weapon on a target range near Moscow. On the field before them were four ZIS-6 trucks. The trucks' rectangular cabs with armored shutters over the front windshields were without bodies on the rear chassis; in its place were slightly inclined tubular constructions rising over the cab. These strange machines were designated as BM-13. They were a special weapon that no army in the world possessed. They were not much at first glance, but a volley of 16 132mm rocket shells blew away a small field of earth.[2]

The "ar-esses" were quickly unloaded and prepared, and they fired their first volley at 3:15 p.m. on 15 July, 1941. The target was the town of Orsha, a staging area west of Smolensk full of German troops and equipment, and the effect was devastating, far beyond expectations. Stunned Red Army troops who fired the weapon thought its explosive discharge "looked like the fire of three full artillery regiments."[3] It caused a corresponding amount of damage - but even greater was its psychological effect. German soldiers in the target area panicked. Russian, as well as German, soldiers in the front lines near the explosions also pulled back in terror. In the interest of secrecy, they had not been warned of the test.

The secret, of course, could be kept no longer. The new weapon was used again that same day to deliver another punishing salvo against the German 12th Panzer Division as it attempted to cross the Oshita River. As the experimental unit positioned itself to strike and waited for the expected tanks, Deputy Front commander General A.I. Eremenko edged forward in order to report on the weapon's performance. He was not disappointed - according to his memoirs, its effect was overwhelming.[4] Within a matter of seconds the simultaneous explosion of 112 rockets broke up the crossing operation, obliterated some 10 tanks and killed more than 100 soldiers. In the words of a shaken German prisoner: "It was a nightmare . . . not only our soldiers at the crossing, but those who were far away were panic-stricken. Everybody ran for his life." It was, he said, "as though hundreds of guns were firing simultaneously."[5]

The career of the fabled Red Army rocket artillery began. A jubilant State Defense Committee, the following day, ordered the mass production of rockets and launchers. The Red Army had disclosed a new and highly destructive weapon - simple to operate, easy to make, reliable and effective in combat.

In a certain sense, of course, there was nothing new about military rocketry. Rockets have a longer history than guns - they had been used by Chinese commanders more than half a millennium before. During the 18th and early 19th centuries, military rockets had been particularly popular in western European armies. A rocket attack is captured in the American national anthem. Though less accurate than slow firing muskets of the time, rockets packed a good deal more punch and enjoyed a much longer range. As the Industrial Revolution improved gunsmithing techniques, however, rockets declined in relative value and were increasingly relegated to a subsidiary role in battle.

Even so, rockets had their military advantages. Their slow launch speed does not subject the warhead to the stress borne by an artillery shell. On a pound for pound basis, rockets can carry more explosive than shells - and a greater variety of explosives. A rocket casing, for example, can easily be notched to fragment on impact. An artillery shell of similar strength, were it manufactured in the same way, would simply blow up as soon as it was fired.

Such factors made rockets attractive to the Red Army High Command, and during the late 1930s, Russian researchers began to test prototypes of a ground-to-ground rocket launching system. Their research program soon resulted in a truck mounted launcher for the RS-132 (hence the acronym 'ar-ess'), an aircraft rocket already in production. The original plans were uncomplicated. The rockets were simply to be fired over the side of the truck that had hauled them to the front. That arrangement proved an unsatisfactory, if expensive experiment, however, and ramps were then fitted longitudinally to the truck chassis.

The change proved successful enough to encourage further experiment, and in 1939, a number of factories in the Moscow area were set to work producing rockets and mounts. At one of them -the Kompressor factory - the design bureau devised a completely new launcher, designated the BM-13.[6] It incorporated the chassis of a six-wheeled, cross-country truck specially equipped with eight long rails. It was capable of launching sixteen 132mm rocket projectiles to a range of some 8,500 meters. Holding two rockets each, the eight rails could be elevated from 15 to 45 degrees on a vertical plane, and could be moved up to 20 degrees back and forth on a traverse plane. The six-foot rockets weighed ninety-two pounds and carried almost eleven pounds of explosive. After a means for electric ignition was perfected, all sixteen rockets could be launched within ten seconds.

Shortly after the German invasion of Russia on 22 June, 1941, the Red Army senior command decided that it would be best to test the new weapon in the field, foregoing the usual testing formalities. Within a matter of days an experimental battery was formed at the First Moscow Red Banner School.[7] Captain Ivan Andreevich Flerov, a recent graduate of the prestigious Dzerzhinskii Artillery Academy, was appointed commander; his battery was equipped with seven BM-13 launching vehicles, and one 122mm howitzer (for the purpose of ranging and registering targets). The unit consisted of forty-four trucks to transport 600 M-13 rockets and 100 howitzer rounds.

After hasty preparation, Captain Ferov's unit left Moscow for the front on the night of 2 July, ready to do its best in the bitter fighting around Smolensk. The battery traveled only at night to preserve its secrecy. At dawn, on 4 July, the battery concentrated in a forest, seven kilometers south of Smolensk, in the 20th Army sector. In the early morning of 14 July, it prepared to leave for Orsha. That morning, Captain Flerov received instructions from the Western Front Artillery staff to strike near Orsha. After completing a short road march to woods east of Orsha, the battery unleashed a volley from the new firing position. And its performance in that engagement was so effective that the unit was used repeatedly as a stopgap in other threatened areas.

On 15 July, the battery in the area of Rudni unleashed three volleys on a German column traveling along Minsk highway. The two encounters were so impressive that by July 22, a second rocket battery had been formed and sent off to the front. This one assisted the beleaguered 19th Army, under the command of Lt. Gen. I.S. Konev. A third battery soon followed.

The name "ar-ess" or "RS" was an acronym for the Russian equivalent of "Rocket Launch System," but because of their effectiveness, the rockets soon received a more affectionate nickname - "Katyusha," a diminutive of Katherine, or in English, "Katie," or "Little Kate." Russians are fond of diminutive nicknames, and no one really knows who christened "Katyusha." The name presumably derives either from the factory mark "K" for "Kompressor" borne on the sides of the trucks that produced the first models. Others, who served with the rocket forces, believe the nickname was from the title of a then popular Russian song at the wartime front.[8] Perhaps both factors played their part.

Not surprisingly, the Germans regarded the weapon with less affection. With the frontline soldier's macabre humor, they called the rockets, "Stalinorgeln," or in English "Stalin's organ," apparently in reference to the rockets' high-pitched shrill as it approached impact.

In August 1941, three batteries of katyushas which had been assigned to the Western and Reserve Fronts were operating. By September three became nine because of their usefulness. The Red Army General Staff formed rocket launcher units and developed their operational concept during the course of the war, based on an analysis of war experiences. Combat experience in the use of the first rocket artillery batteries confirmed significant advantages: maneuverability, potential to easily shift launchers to threatened sectors, massing characteristic of rocket fire, distinctly high effectiveness, especially destruction of discovered enemy force locations and showed a strong demoralizing impact on enemy soldiers.[9]

The Stavka, or Soviet High Command, had ordered a radical increase in production of rockets and launchers. Furthermore, it decided to form eight new "Guards Mortar Regiments," exclusively devoted to rocketry. The title was a designation that would continue to protect the true nature of the new weapon, and offer some deception in the Red Army order of battle. In a matter of weeks, rockets became an integral part of the Red Army.

The first eight new regiments were formed in September; others followed in short order. Each consisted of three battalions, composed, in turn, of three batteries each. Each battery operated four launchers, either M-13 or M-8. In addition to rocketry, at least, some new regiments enjoyed the services of a battery of antiaircraft protection and a detachment of armored cars. All included a maintenance unit. They came to the front well-equipped for their task. And, in a single volley, each powerful regiment could deliver 576 M-13 rockets or 1296 M-8 rockets - nearly twenty-four tons of deadly metal.[10]

Additionally, in the month of September, operational groups for the Guards Mortar Units were created in the Front staffs of the Northwestern, Southwestern, Southern, and Bryansk. A little later, operational groups were at the Western, Kalinin, and Volkhov Fronts. The Front operational groups supported assigned rocket units. Their support actions were exemplified in such actions as securing a resupply of rockets for guard mortar units running low in rockets during the course of an operation.

In the spring of 1942, a captured German sergeant testified to the overwhelming power of such a volley. Severely wounded, he had been picked up the previous autumn after a rocket strike in the area of Dikanka, west of Kharkhov - one of the first attacks conducted by the new Guards mortar units. The 4th Regiment was, in fact, the first such Katyusha force to arrive at the front, and at Dikanka had worked in support of the Red Army 14th Cavalry Division.

Still shell-shocked, the sergeant remembered "there was the instant presence of scorched earth . . . massive fire! The fire devoured everything: our soldiers, guns, armored vehicles, tanks . . . We lost then many soldiers and officers." The sergeant related that among the wounded in the hospital, there was a rumor the Russians had struck with a secret, super powerful shell from a multitude of guns.[11] Rocket-fire was all the more terrifying because German soldiers were unsure just what they were up against.

In the summer and autumn of 1941, German commanders wanted more information about the new weapon and often sent patrols to find katyusha units behind the lines. SS General Max Simon, in a paper which he wrote on his experiences against the Russians, remarked that some trucks with the simple rocket launch rails had been captured but that the significance of the frame had not been appreciated for some time.[12]

In other cases, German Army units made detailed studies, as intelligence reports to be sent to higher commands, to understand the new weapons. In December 1941, for example, the XXXI Army Corps, First Panzer Army provided a report with detailed sketches and a series of photographs of the rockets.

Meanwhile, Soviet research and development continued. By September, the BM-8 launcher was available in two versions, the first truck-mounted and able to fire simultaneously thirty-six rounds of 82mm rockets, the second mounted on a T-60 light tank chassis and able to fire twenty-four rounds. The smaller version of the launcher soon proved more valuable. It was inferior to its truck mounted counterpart in firepower, of course, but far superior in cross-country mobility. The tracked launchers could offer much closer support to mechanized and tank units.

The light BM-8s proved especially useful in the later years of the war, when the Red Army had seized the offensive. They were assigned to advance parties and mobile groups ranging far ahead of the main Red Army forces as fire support, in order to disrupt German rear areas and prevent the forward deployment of enemy reserves.

In the early war years, Katyushas were most often used on defense - they proved particularly effective in reinforcing threatened rifle divisions. A detachment of Katyushas dramatically increased a division's firepower, and so helped to stabilize a defensive front. In those sectors where few katyusha units were allotted, Red Army commanders employed them at the most critical moments on major avenues of attack by the Germans.

In September, the 42nd Separate Guards Mortar Battalion was formed, and it included Captain Flerov's battery. Katyushas became involved heavily in difficult, close fighting and began to take their share of casualties. On 6 October, 1941, Captain Flerov's battery, cut off in the vicinity of Vyazma, attempted a breakout from encirclement. Hoping to reach their own lines, Flerov led the battery through a forest. Upon emerging from the woodline, several German panzers appeared, trapping the battery under heavy fire. After firing a salvo at point blank range, Katyusha crews, following Captain Flerov's lead, blew up all launchers to preclude their falling into German hands for close inspection. Some forty-six men of the battery broke out of encirclement. Captian Flerov was mortally wounded.

Despite their potency in combat, however, by November 1941, a change in katyusha tactics seemed necessary. Newly formed rocket regiments proved cumbersome, too large for inexperienced commanders to control without radio help. Furthermore, because new regiments lacked adequate reconnaissance capabilities, they were frequently assigned missions that wasted their overwhelming firepower. More frequently still, their component units were employed by sector commanders piecemeal, wherever the need appeared greatest. Such pressures persuaded the Red High Command to reorganize nine of fourteen guards mortar regiments into twenty-eight independent rocket battalions.[13] These battalions were smaller and more manageable units which had proved consistently the most valuable in combat. Each battalion contained two batteries of eight launchers apiece.

By late fall, thirty-three of fifty-nine katyusha battalions in service were concentrated in the Moscow defense zone. Leningrad Front had five battalions, Southwestern Front nine, Southern Front six, and other Fronts one or two battalions each.[14] German commanders complained in their reports of the devastating effect on morale as Katyusha rockets were employed increasingly in greater numbers.

When the battle for Moscow climaxed in the winter of 1941-42, the importance of Katyusha units grew. The Red Army began using five to ten battalions at once in a major sector, but such deployment led to difficulties in fire control and supply. Under these conditions, an array of small, independent Katyusha units merely decreased efficiency. Many previously separate battalions were reconsolidated into larger formations. In January, the Stavka decided that it would be best to form twenty new Guards Mortar Regiments.

The Stavka Supreme High Command, in a directing letter dated 10 January, 1942, generalized the Red Army's experiences in counteroffensives around Moscow, analyzing shortcomings in actions of forces, and in particular, the use of artillery. The directive pointed out insufficient artillery support to infantry and armor in the course of all the offensive battles. The directive became the basis for new tactical methods and organization of artillery operations. Guards mortar regiment organization consisted of three battalions of two batteries of four launchers, BM-13 or BM-8. A volley of an M-13 regiment launched 384 rockets; BM-8, 864.[15]

The Red Army wanted to follow its successful winter operations by maintaining the strategic initiative with a spring offensive. The Red Army High Command calculated an offensive around Kharkov that could be conducted without encountering significant German Army reserves. On 12 May, 1942, anticipating a German offensive by five days, Marshal S.K. Timoshenko, commander of the Southwestern Front, launched a powerful "decisive offensive," as he called it in his order of the day. The Southwestern Front's operation sought to envelop German forces in the Kharkov area from north to south, carrying the offensive farther to the west and the Dnepr river.

With faulty intelligence on the size and location of the German buildup on his flanks, Timoshenko, on the morning of May 17, decided to commit the 21st Tank Corps to the deep drive on Kharkov. That day Army Group Kleist, including one panzergrenadier, two panzer and eight infantry divisions, crashed into the southern flank of Timoshenko's forces. Rapidly advancing German forces imperiled the entire Southwestern front offensive at the base of its salient.

By the next morning, the Red Army's inability to hold the German onslaught became evident to the Soviet field commanders. But Stalin refused to halt the offensive. By evening, German units, advancing to the north 40 to 50 kilometers, reached the Northern Donets River. Army Group Kleist's advance along the west bank of the Northern Donets created an immediate threat for the entire force operating in the Barvenkovo salient.

Not until the evening of 19 May, when two armies and a special cavalry group were in certain danger of being surrounded, did Timoshenko order a halt to the advance. Stalin agreed with the order. But, the decision came too late. As Army Group Kleist's spearhead struck north to Balakleya and Izyum, the German Sixth Army attacked south across the Northern Donets. The two forces met and turned westward, waiting for Timoshenko's breakout attempt.

Withdrawal routes across the Northern Donets were cut off for Red Army troops remaining in the salient. In fights of appalling slaughter, groups of Russians infantry and tanks charged desperately against the closed German ring. Isolated groups of Red Army troops succeeded in slipping out of the encirclement and crossing to the eastern bank. In a battle of annihilation, the Germans claimed destruction of fifteen infantry divisions, seven cavalry divisions, ten tank brigades, and the capture of 200,000 troops. In the trapped force were two Guards Mortar Regiments, 5th and 55th, supporting the offensive.

Like the reports previously noted from the XXXI Army Corps, a prisoner of war report revealed information to the operation of the Katyusha system. Lieutenant Vladimir Kolbasov, 55th Guards Mortar Regiment, who was trapped in the failed offensive, described the rocket system in detail to his captors. In his unit, the M-13 was mounted on the ZIS-6, International, Studebaker, and Ford trucks. The American trucks, as part of the Lend-Lease program, came across Siberia to Gorki where they were transformed into Katyusha launchers. The eight double launch rails could fire sixteen rockets in thirty-five seconds. The electrical firing mechanism consisted of two nine-cell batteries each eighteen volts and eighty ampheres. The design for series firing was adjusted so that rockets alternated firing left and right off the rails to prevent overloading one side, disrupting accuracy and tipping over the launcher vehicle.[16]

In fighting the system, katyushas normally stayed in hide positions. Only when the systems were required the weapons would move forward to a loading position, and only when the target is clarified or designated, are the systems brought forward to one of three firing positions. These positions were located 300-400 meters forward of loading positions. The Katyushas unleashed their salvo and returned to the loading position. Katyushas remained only five or six minutes in the firing position. The return trip, reload, and return to firing position required fifteen minutes.

The new year (1942) saw changes in equipment as well. In June, Soviet designers developed a pair of more powerful rockets, the M-20 and the M-30. The M-20, designed to be fired from already existing BM-13s, was 132mm in caliber, weighed in at 127 pounds and had a range of 5,000 meters. Because of size, however, the new rocket could be fired only from upper launcher rails. An M-20s volley would consist of eight rockets, compared with sixteen M-13s. The new weapon soon revealed other weaknesses as well. Field use quickly demonstrated that the M-20 lacked explosive and fragmentation power, and eight rockets per volley proved insufficient to attain desired intensities of fire. By 1943, production of the M-20 ceased.

The new M-30 fared better. It was a larger and bulkier rocket, 300mm and 158 pounds, and its range was correspondingly shorter, only 2,800 meters. The M-30 required a new launching apparatus, a special launching frame set on the ground, which could handle only four rockets at once packed in special sealing. Despite these drawbacks, the M-30 enjoyed one overriding virtue - it fragmented to deadly perfection.

Even so, it took time to learn how to handle the new weapons effectively. Initially, M-30 units were formed into separate battalions, and the first volley of heavy rocket artillery was fired on 5 July, 1942, in the Tul'sk district of the Western Front. Results were encouraging, but further combat use revealed shortcomings. Because each battalion required ninety-six launch frames, each of which had to be set up on the site, it took time to select and arrange firing positions. The new units were nowhere near as mobile as their established light-rocket counterparts. Ammunition transport was also a problem. Trucks assigned to the heavy battalions were sufficient to carry only half a volley at one time, a factor which further decreased mobility. Additionally, the new rockets threw up so much smoke and dust when launched that their emplacements were an easy target for enemy counter fire. Their mission accomplished, heavy units had to evacuate both equipment and soldiers quickly - something they found difficult to do.

Transport shortcomings, inability to provide adequate communications, and a lack of reconnaissance capability forced the abandonment of heavy katyusha regiments. The first five M-30 regiments were disbanded and their battalions became independent. In the immediate reorganization, separate M-30 battalions were reorganized with two batteries possessing forty-eight launch frames each.[17]

Simultaneously, in 1942, with the development of heavy rocket artillery, the Guards Mortar Units with BM-13s and BM-8s continued to grow impressively. In January 1942, eight regiments and seventy-three separate battalions were operational. By 1 October, 1942, the Red Army had seventy-nine M-8 and M-13 regiments, seventy-seven separate M-30 battalions, thirty-six separate M-8 and M-13 battalions for a total of 350 battalions.[18]

In the battle for Stalingrad, rocket artillery units were far more numerous than they had been in the battle for Moscow the previous winter. As a general rule, katyusha regiments supported one to three divisions each, but rifle divisions in crucial sectors received many times that much support, one or two regiments per division. Army commanders had separate battalions and regiments in their reserves. In the battle along the Volga, the Russians used high Volga river banks as garages for katyusha vehicle launchers. Trucks backed out of their cavern garages, fired a salvo, and moved back under cover in a matter of a few minutes.[19] From August to November 1942, Guards Mortar Units in the Stalingrad Front alone fired more than four million rockets.[20]

Guards Mortar Units maneuvered, volleyed, again maneuvered, and again volleyed. Katyushas inflicted strikes on German forces in their assembly areas or on avenues of advance, and they fired in direct fire mode to repel massed infantry and panzer attacks.The Guards Mortar Units were used extensively in the counteroffensive. German units, as a rule, counterattacked as Red Army forces penetrated the forward edge of the main defensive zone. They immediately committed to battle their battalion, regiment, division reserves. Guards Mortar Units played a decisive role in repelling German counterattacks.[21] A quick response would separate infantry from panzers, then panzers could be dealt with using other assets; such as self-propelled assault guns and artillery indirect fire.

As time passed, the Red Army grew more and more used to its new heavy weapons, and as the M-30s were more skillfully deployed, many problems which first crippled large rocket units were overcome. The Stavka accordingly decided, in December 1942, to form Heavy Guards Mortar Divisions, designed, at first, to consist of four light regiments of M-13s and two brigades of M-30s. A volley from a division consisted of 3,840 rounds, representing a throw weight of nearly 230 tons of explosives.

During the winter of 1942-43, the Red Army shifted strategically from defense to offense. The varied katyusha units accordingly had to learn methods of employing their withering firepower in the attack. The task, once again, was especially difficult for heavy brigades. They proved invaluable in the preceding artillery preparation against defenders and in supporting the initial phase of an attack, but they could not easily follow tanks and infantry as they penetrated the depth of a German position.

The transport problem for larger Guards Mortar Divisions was alleviated by American lend-lease trucks. Experience in the 5th Guards Mortar Division reveals that American trucks were a mixed asset.[22] On 18 January, 1943, the division received its remaining motor transports with American machines marked, "Chevrolet," "Ford," and "Dodge." Neither drivers nor officers knew how to operate the American trucks or understood their characteristics. Drivers, mechanics, and commanders of units spent all day the 18th and all night the 19th familiarizing themselves with the trucks and preparing them for the coming road march into battle. Preparation became a difficult affair. In freezing weather, the truck engines started poorly. Drivers had to put heated water from marching kitchens into radiators. Once on the road it was hard going in the winter months, especially for units with American trucks. On fine roads they moved with great speed, but when it began to snow heavily, they became stuck and required extensive towing.

But new divisions had shown their worth in spite of all difficulties, and early in 1943, the 5th and 6th Heavy Guards Mortar Divisions were formed; in February, the 7th. These new divisions, each with three M-30 brigades, were capable of launching 1,152 rockets per volley. Each battalion had seventy-two launch frames that could deliver 288 rockets.

In 1943, rocket divisions grew steadily in strength and all the more so after the development of the M-31. Like the M-30, the M-31 was a 300mm rocket, but it had a greater range, 4,325 meters to 2,800 meters. The improved launch frame for the M-31 allowed for eight rockets rather than four, doubling in a single volley the potential firepower of heavy units. The M-31 enabled a brigade to fill a target area of forty-two hectares (roughly 103 acres) with some eighteen rockets per hectare, a density of fire sufficient to put seven or eight rockets at once on an area the size of a football field.[23]

The military councils and staffs of the Guards Mortar Units around Stalingrad systematically collected and analyzed information about the combat use of rocket artillery, studying and generalizing experience of maneuver units, creating Guards mortar Unit groups, massing fires, and other actions. The results of this massive effort created the "Instructions for the Use of Mortar Units M-30, M-20, M-13 for the Breakthrough of Reinforced Enemy Sectors." This document, adopted by the Defense Commissar on 4 April, 1943, became the regulatory guidance for use of rocket artillery. In particular, it contained requirements for compulsory inclusion of Guards Mortar Unit volleys in the general scheme of artillery fire in offensive operations. This was the single most important lesson, extracted from experience in the Stalingrad battles.[24]

Experience showed that the number of tube artillery for a breakthrough operation was extensive in density per kilometer of front, and the coordination of such a high number of weapons proved slow and cumbersome when compared to the guard mortar units. In one study, twelve regiments of M-31 and three brigades of M-31 replaced thirty-eight regiments of tube artillery.[25] The M-31 brigade by staff calculations equalled seven to eight regiments of corps artillery in comparable destruction.[26] In a corps breakthrough sector with fortified sections, no less than two M-30-31 brigades were planned.

In the extensive battle of Kursk, in July 1943, Red Army forces put to successful use approximately 496 BM-8 and BM-13 combat launchers, and 2,626 heavy M-30 and M-31 launchers.[27] The allocation to defending Fronts on July 4 was Central Front had ten regiments, one division, and one separate battalion, and the Voronezh, eleven regiments and one separate battalion. Commander of the Guards Mortar Units who reported to Commander of Red Army Artillery and Stavka, underscored that the density of rocket artillery on one kilometer of front in the Kursk salient was two to three times higher than in the battle for Stalingrad.[28]

By late 1943, the Red Army High Command could call upon services of seven rocket divisions, thirteen brigades of M-31s, ninety-six regiments of M-13s, nine regiments of M-8s, thirty battalions of M13s and nine battalions of M-8s. At this point in the war, the army generally could place on main strike sectors two to four regiments of M-8 and M-13, along with one to three brigades of M-31.[29] In two years the Red Army rocket artillery had come a long way as an integrated fighting capability on the battlefield.

In 1944, Soviet designers worked out improved versions of both the M-13 and M-31. The range of the new M-13UK was nearly 7900 meters; M-31UK, 4,000 meters. A Guards Mortar Brigade of M-31UK could work over an area with a front of 900 meters and to a depth of 220 meters. The M-31 rocket made a crater seven to eight meters in diameter and two to three meters deep.[30] These heavy brigades would have the mission to knock out the first line of defense on the forward edge eliminating centers of resistance, strongpoints, and other engineer obstacles which were located in a particular sector.

By 1944, Katyusha units were well along the way towards refining their tactics for offensive combat. Rocket units in the attack could fire several salvoes during the artillery preparation phase into various sectors. During the assault they could provide supporting fires to advancing Red Army troops to the limits of their range. Before the attack, firing positions were prepared and occupied. Enough ammunition for the planned number of salvoes was brought forward. Observation posts were established forward between firing positions and the German front lines.

Guards Mortar Units in the hands of combined-arms commanders served as a powerful offensive weapon. In the offensive the katyusha brigades fulfilled the following missions:

- destruction of fortified enemy centers of resistance
- destroy soldiers and equipment on the forward edge
- destroy population centers incorporated in the enemy defense.[31]

German reports indicated that the rocket forces were employed in different sectors of the front. Their harassing fires were directed mainly at the forward line and immediate heights. However, occasionally rocket salvoes supported Red Army patrolling and reconnaissance. For first fires against the forward edge, katyusha units attempted to have a minimum of twenty to thirty rockets per hectare.

As the Red Army attack advanced into the depths of the German defense, katyusha units would advance with it, moving forward three or four kilometers at a time, and firing from each new position gained. To advance, prepare, and fire took a rocket crew about thirty minutes, and such rapid movement was essential for work close to German lines. By this time, German responses to rocket attacks had become very swift indeed, and in order to survive, Katyusha units had to fire and move on at once.

Targets for Katyushas in the depth of German defenses were selected to repel infantry and tank counterattacks, protect the maneuver reserves, neutralize strongpoints on intermediate, or secondary lines of defense, and other important objectives. Massing fires on large concentration of tanks, the M-30 and M-31 rockets within five to ten meters could destroy light or medium tanks.

In spite of improvements in offensive tactics, mobility remained a problem for katyusha units. By 1944, rocket forces were assisted by motorcycle reconnaissance teams, but this did not help to move the weapon launchers. Withdrawing German units often destroyed roads and bridges as they retreated, and this made it extremely difficult for the rocket artillery to advance. Special road construction teams were raised to help them across not only German obstacles, but also marshy or wooded terrain. As always, the Red Army was short on equipment by Western standards, and was forced to make do with ingenuity and muscle. Unit personnel had to do some road construction with entrenching tools, use towing cables and chains to assist transport vehicles over difficult, broken terrain.

In the spring of 1944, the BM-31-12 was introduced to overcome these terrain difficulties. This launcher, as a combat vehicle, could move over broken terrain, and allowed Red Army heavy rocket artillery to accompany tanks and infantry in all phases of an offensive. Each brigade was assigned one battalion of these new launchers, each of which could fire twelve rockets after only seven to thirteen minutes of preparation. The mobility of BM-31-12 launchers allowed them to suddenly mass destructive fires on strongpoints and German pockets of resistance.[32] The soldiers, understandably, loved them, and nicknamed them "Andryusha" and "Vanyusha" or, in English, "Andy" and "Johnny" - helpful big brothers to "Katie."

With the development of "Andy" and "Johnny," Red Army rocket arsenal reached its final shape of World War II, and the new weapon played a major role in the final campaigns. At the beginning of the Berlin operation in April 1945, the 1st Ukrainian, 1st and 2nd Belorussian Red Army Fronts had at their disposal 1,519 launcher vehicles and 1,785 launching frames. At one point, they simultaneously fired nearly 30,000 rockets - it would have taken 840 regiments of conventional artillery to match a barrage. Each rifle division in the first echelon had three to four battalions of M-8, M-13, or M-31-12. On a one kilometer section in a breakthrough sector, the density of launchers was significantly increased. For example, in the 5th Shock Army, 1st Belorussian Front were thirty-eight launcher to the kilometer; 49th Army, 2nd Belorussian Front, twenty-seven launchers.[33]

On 1 May, 1945, the Guards Mortar Units had seven battalions, eleven separate brigades, 114 separate regiments, thirty-eight separate brigades, and totalled 3,081 combat launcher vehicles.[34] The Katyusha units became an important aspect of Red Army artillery operations contributing to its combat traditions. Over the course of four years, "Little Kate" had thoroughly proved her worth. The Guard Mortar Units became the basis for the creation of the Red Army Strategic Rocket Forces, as well as establishing a legacy for an elite force of the "Boebaya Machina" in Soviet Army artillery.

NOTES

1. A.I. Eremenko, *The Arduous Beginning*, Moscow: Progress Publishers, 1966, p. 163.
2. P.A. Degtyarev, P.P. Ionov, *Katyushi na polye boya*, Moscow: Voenizdat, 1991, p. 4.
3. N.M. Afanasev, *Pervye zalpy*, Moscow: Voenizdat, 1982, p. 44.
4. Eremenko, p. 163.
5. V. Ryabchlikov, *The First Rocket Battery*, Soviet Military Review, Vol. 11-1978, p. 13.
6. P. Degtyarev and V. Gurkin, "Stanovlenie i razvitie reaktivnoi artillerii v pervom periode voiny," *Voenno istoricheski zhurnal (Vizh)*, Vol. 12-1976, p. 24.
7. Ibid.
8. Degtyarev, Katyushi, p. 17.
9. Ibid., p. 13.
10. Ibid., p. 14.
11. *About the Legendary Katyusha*, Kiev: Izdatel'stvo Tsk LKSMU "Molod," 1975, pp. 60-1.
12. James Lucas, *War On The Eastern Front 1941-1945*, New York: Bonanza Books, 1979, p. 47.
13. I. Prokhorokov and V. Trusov, "Reaktivnaya artilleriya v Velikoi Otechestrennoi Voine," *Voenno istorischeski zhurnal*, VOl. 1-66, p. 4.
14. Ibid.
15. Degtyarev, *Katyushi*, p. 46.
16. Oberkommando 17, Interrogation Report, dated June 16, 1942.
17. Degtyarev, *Vizh*, p. 31.
18. Degtyarev, *Katyushi*, p. 54.
19. Alan Clark, Barbarossa: *The Russian-German Conflict 1941-45*, New York: William Morrow and Company, 1965, p. 224.
20. Degtyarev, *Katyushi*, p. 75.
21. Red Army General Staff, *Sbornik takticheskikh primerov po opytu otechestvennoi voiny No. 13*, March-April 1945, p.6.
22. V.V. Gurkin and A.E. Ivashchenko, *5-ya gvareiskaya Kalinkavichskaya*, Moscow: Voenizdat, 1979, p. 12.
23. Prokhorokov, p. 5.
24. Degtyarev, *Katyushi*, p. 93.
25. Red Army General Staff, *Sbornik takticheskikh primerov po opytu otechestvennoi*

voiny No. 12, January-February 1945, p. 75.

26. Ibid., p. 81.

27. *Legendary katyusha,* p. 31.

28. Degtyarev, *Katyushi,* p. 107.

29. Ibid., p. 148.

30. *Sbornik No. 12,* p. 83.

31. Ibid.

32. *Legendary katyusha,* p. 30.

33. Degtyarev, *Katyushi,* p. 227.

34. "Gvardeiskie minometnye chasti," *Soviet Voennaya Entsiklopediya,* Vol. 2, Moscow: Voenizdat, 1976, p. 493.

AUTHOR'S NOTE: I dedicate this chapter to my friend, Tibor Ruff, who as a young boy watched Katyushas fire in the Red Army assault on Budapest.

CHAPTER TWO

The Bukrin Drop: Limits to Creativity

On the evening of 24 September, 1943, Red Army paratroopers filed across the tarmac of two airfields to board transport aircraft. The Soviet High Command had identified an opportunity to preempt a German Army defense along the wide Dnepr river south of Kiev. In an innovative use of combat forces, an airborne operation was to assist Red Army units crossing the river to recover agriculturally and industrially rich lands in the western Soviet Union. The battle plan for what became the Red Army's single largest airborne landing in the Second World War was to create a protective perimeter on the western and southwestern approaches to the initial Red Army bridgeheads in the Bukrin bend of the Dnepr river. However, within twelve hours of the first wave's departure, the operation was terminated. Paratroopers were dropped everywhere along the Dnepr except in mass on the intended drop zones. With no unit cohesion, the widely dispersed paratroopers were easily eliminated by the German units. The Bukrin drop was a failure.

Official Soviet histories of the operation are few and incomplete.[1] Other accounts have failed to balance the few and partial Soviet sources with German unit operational reports. Conventional accounts rely instead on sometimes inaccurate postwar recollections by German army officers.[2] The present study extends the inquiry to include a full range of available sources, thereby resolving some of the gaps and uncertainties in past accounts for a better understanding of how this perfect opportunity for airborne employment slipped to disaster.

Examining this slippage suggests creative limits to initiative in combat. Creativity and initiative are interdependent and, as long as the elements of these highly prized attributes are not clearly recognized in all their parts, our understanding of past events remains incomplete. Creativity is based, in part, upon a thorough knowledge and understanding of friendly troops capabilities, their armament and technology, and methods of controlling troops. Just as master artists had to have the facility and dexterity to manipulate physical substances, paints, brushes, to incredible levels of creativity, so too masters in the art of war must have command of basic substances to demonstrate creativity and initiative on the battlefield.

After the Red Army halted major German attacks at Kursk in July 1943, German forces spent most of August fighting to hold back a Red Army counteroffensive. German commanders had difficulty coping simultaneously with Soviet advances and Hitler's reluctance to give ground. Field Marshal Erich von Manstein, commander of Army Group South, believed Red Army leadership was planning a major push for the Dnepr river, and he attempted to persuade Hitler that either his southern wing be reinforced with additional forces or he be given freedom of movement to halt the Red Army on a shorter line farther back.[3] Hitler did not respond.

The Dnepr river with its higher west bank and broad expanse offered a significant obstacle to a continued Red Army advance. Although Hitler had forbidden extensive engineer preparations on the Dnepr for fear of inducing a quicker retreat by his army, the German commander believed a strong, stabilized defensive line could be held along the river. The Soviet command, on the other hand, wanted to seize bridgeheads over the Dnepr to maintain fluidity and momentum in their offensive drive.

On 14 September, General Nikolai Vatutin's Voronezh Front shattered von Manstein's northern wing and pushed within seventy-five miles of Cherkassy on the Dnepr. Hitler reluctantly consented to retreat, and the race for the Dnepr river crossings began.

For von Manstein's army group, the problem of moving three armies behind the Dnepr river was a formidable operational task. From a front 440 miles wide, approximately one million men with equipment had to converge on a maximum of five Dnepr crossings. After the armies crossed, they had to fan out to cover the unprotected west bank. Red Army leadership correctly assessed this moment of extreme German vulnerability. They sought to seize this temporary opportunity by a forced river crossing without pause and an airborne operation.

Use of airborne troops in conjunction with ground troops contributes to a creative employment of forces on the battlefield, allowing the commander to achieve surprise and show initiative. Although airborne forces offer swift operational employment, an airborne operation requires extensive lead time in planning and preparation, such as location of drop zones in target area, coordination of aircraft support, assembly and loading of forces, weather

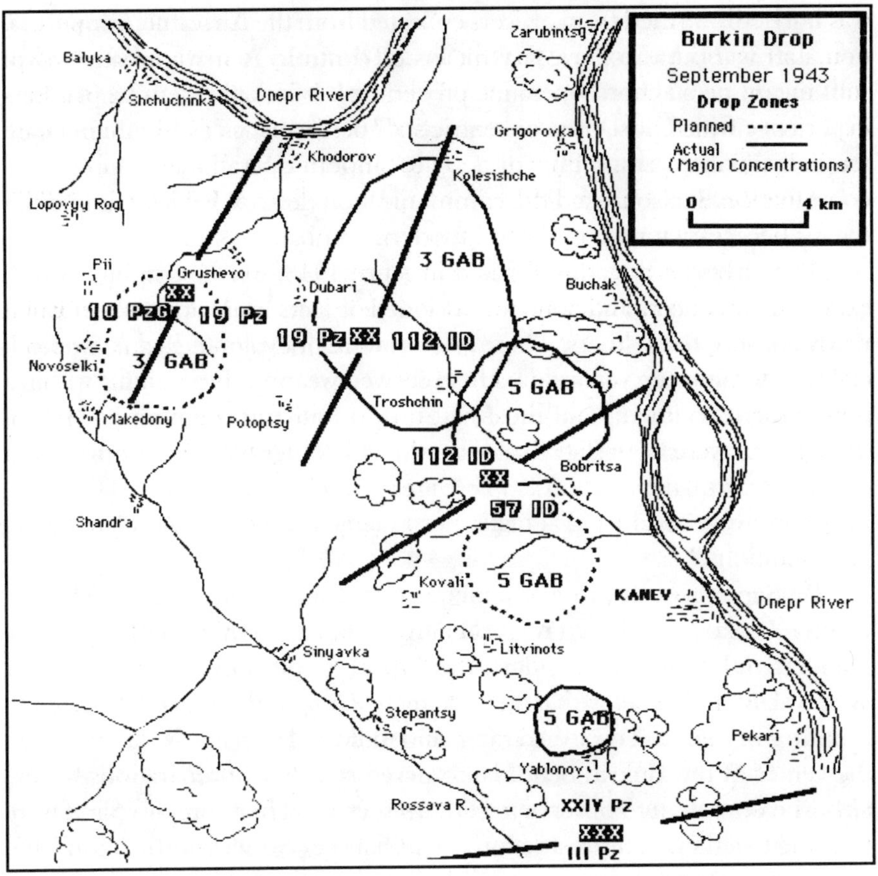

conditions, and ascertaining the enemy situation. All this assumes well trained and qualified forces. Anything short of careful deliberate planning and coordinated execution invites trouble.

The Soviet Headquarters, Stavka, had planned ahead. In late summer 1943, the Stavka ordered Major General A.G. Kapitokokin, commander of airborne forces, to intensify the training of large units for seizure of bridgeheads on major water barriers. In August 1943, a large tactical exercise conducted on the banks of the Moskva river near Moscow involved the landing of the 3rd Airborne Brigade.[45]

By early September, as the Red Army offensive towards the Dnepr gained ground, the 1st, 3rd, and 5th Guards Airborne Brigades (GAB) were detached from their respective airborne units to form the newly-designated VIII Airborne Corps. The deputy commander of the airborne forces, Major General I.I. Zatevakhin, was named corps commander. The corps staff was hurriedly formed from officers collected from the Airborne Administration staff without the necessary means of control.[5] A newly-created corps staff meant no standard operating procedures. It lacked extensive planning experience which develops concentric staff performance rather than the often individual, eccentric impediments to smooth operational planning and coordination. It also lacked the communication devices, vehicles, and other means necessary for troop control modern combat.

The airborne brigades, formed in April 1943, were a combination of experienced officers and noncommissioned officers from airborne units and newly conscripted soldiers. These recruits were not volunteers and averaged eighteen to nineteen years old.[6] They received weapons fire training within a replacement regiment, and most had two to four training jumps.[7] Half of these jumps were from static balloons while the other half were from aircraft over their permanent stations.[8] There are no indications that any of the training jumps were conducted at night, nor as large units under simulated combat conditions.

By September 16, in a key meeting held in the presence of Marshal Georgi Zhukov, the Stavka representative, Nikolai Vatutin, the Front Commander, and other commanders, the airborne operation was reviewed for approval. Standing before a large scale map, Zhukov announced the need to capture a bridgehead over the Dnepr and indicated a relatively flat surface in the center of the Bukrin bend. He ordered sternly all preparations for the airborne operation to observe maximum secrecy. Therefore, no plans were to come together.[9] This restriction would hamper Soviet coordination later.

Zhukov raised the question of when to drop the airborne forces -day or night. Aviation General Nikolai Skripko, commander of the Long Range Aviation (LRA), reported that transport crews in the past few years had flown exclusively at night, as a countermeasure against German fighters. Many of the pilots had flown at night and in difficult weather, but they never flew in formations. Skripko opined that his crews, thrown together in sections and squadrons, would need four to five days of training for daytime formation flying. Additionally, there were not enough Red Air Force fighters to surround and protect the transport formation. Zhukov decided to conduct the drop at night.

Other problems included a shortage of fuel and fuel trucks at the staging airfields. General Skripko, again bearing bad news, reported that his latest information indicated that fuel trucks were on railroads heading for the front and not at the airfields. There was no gasoline anywhere that day.

"To me you are hampering the landing," Zhukov snapped. But he promised to take corrective measures.[10]

Marshal Zhukov stressed the need for coordination between the Voronezh Front and airborne forces in the operation, and he ordered General Vatutin to update the airborne brigades' plans. Available Front artillery was directed to concentrate fires south of the landing zones to inhibit German units moving into the landing area.

Airborne units would have two missions. First, they were to seize and secure defensive positions in the base of the Bukrin bend before the enemy could organize a firm defense along the west bank of the Dnepr river. Second, while consolidating for future operations, paratroopers were to prevent the movement of German reserves against the bridgeheads of the Voronezh Front. Ground forces of the Front were to link up with the airborne brigades within three days.

Meanwhile, on 18 September, the XXIV Panzer (Pz) Corps, complying with Field Marshal Erich von Manstein, commander Army Group South, directive to fall back to the Dnepr, was ordered to cross at Kanev. The movement to Kanev created a breach approximately 125 kilometers wide (from Kanev to Kiev) between the two German Armies crossings over the Dnepr. The 3rd Guards Tank Army (GTA) entered the gap in a race with XXIV Pz Corps to the river.

The 3rd GTA reached the Dnepr river opposite the Bukrin bend by nightfall on 21 September. The Army commander, General Pavel Rybalko, ordered his advanced battalions to cross the river without waiting for bridging or artillery support. With rafts and small boats, his lead unit grabbed the west bank in the vicinity of Grigorovka.[11] In the first hours, because the 800-yard-wide Dnepr hindered rapid buildup, only small Red Army units with no artillery or tanks held the bridgehead.

German commanders, knowing once the Red Army succeeded in establishing a bridgehead it would be nearly impossible to reduce, responded quickly. While waiting for the XXIV Pz Corps to cross and occupy the unprotected west bank on 21 September, the Fourth Panzer Army to the north dispatched the 19th Panzer Division (PD) to the Bukrin bend. Since the division itself was crossing, only the reconnaissance battalion under the command of Major Guderian, the youngest son of the famed panzer leader, could respond. On 22 September, the division's 73rd Regiment followed to occupy positions at Balyka and in the Bukrin bend.

By midday 22 September, pressure from Soviet crossings at Grigorovka and Zarubintsy increased XXIV Pz Corps commander's concern for reinforcing as soon as possible the 19th PD with newly-assigned units.[12] However, these units, 7th Panzer and 72nd Infantry Divisions, would take too long to arrive.

On 23 September, the XXIV Pz Corps was still on the east bank and vulnerable. While corps elements began crossing a shored-up railroad bridge at Kanev, 3rd GTA units attacked the bridge. Tenacious fighting by the German rear guards allowed the corps to cross unimpeded. By 2215 hours, 57th Infantry Division (ID) had crossed the river to positions in the vicinity of Kanev. The 112th ID, in the process of crossing, hurriedly sent its lead elements to reinforce 9th PD near Grigorovka.

Meanwhile, the Voronezh Front slowly expanded its bridgehead in the Bukrin bend. General Kiril Moskalenko's 40th Rifle Army made additional crossings west of the Bukrin at Balyka and Shchuchinka. Since the planned airborne landing lay in 40th Army's zone of advance, General Vatutin arrived at Moskalenko's command post on the morning of 23 September. The situation in the designated drop zones was not clear. An autumn weather front with heavy rains and high winds prevented aerial reconnaissance obtaining exact German unit dispositions and other information. The Front commander's desire for more information delayed approval of the airborne force's mission until midday on the 23rd.

But the moment seemed right for the airborne drop. Planned concentration of troops and materiel at the embarkation airfields was to have been accomplished by 20/21 September, two days before the drop. The 5th GAB moved by rail in two trains and transported one battalion by trucks to Bogodukhov airfield, northwest of Kharkov.[13] Beginning on 19 September, 3rd GAB was transported in three trains by way of Kharkov to Lebedin, south of Sumy.[14] The brigades' new 45mm field guns were transported separately to a third airfield, Smorodina. Because retreating German forces had destroyed rail lines and rolling stock, heavy demands on recently-restored railroad tracks and insufficient rolling stock delayed the brigades' arrival at the airfields until 24 September. The new staff had not taken into account these wartime frictions on plans.

Train delays forced postponement of the operation for one day and created other complications. Originally, 1st and 5th GABs were to make the initial assault, but the 1st GAB did not complete its concentration at the airfield in time. The VIII Airborne Corps staff decided to have the 3rd GAB at Lebedin go in its place.[15] The 1st GAB became the corps reserve in preparation for a drop on the second or third night of the operation.

Swapping unit missions at the last moment substantially accelerated the sequence of planning and preparation required by 3rd GAB commanders and their staffs. All day on 23 September, 3rd GAB planners spent their time studying the new mission, appraising the situation, making decisions, and developing load plans. The final decisions, delayed by the Voronezh Front, were not made until the following day, 24 September. The brigade commanders did not receive the final parts of their orders until an hour before paratroopers loaded into planes.[16]

At 0500 hours, 24 September, the XXIV Pz Corps' rear guard commander ordered the Kanev railroad bridge destroyed. That accomplished, and without respite, the 10th Panzer Grenadier Division (PGD) quickly moved to its sector opposite the 40th Army crossings at Balyka. By dawn on that day, the absence of German units on the east bank undoubtedly alerted the Voronezh Front that XXIV Pz Corps had withdrawn completely to the west bank. However, Red Army activities for the airborne drop progressed.

Problems in concentrating the Soviet forces continued. For the airborne landing 180 Li-2, transports, and 10 Il-4 to tow 35 A7 and G11 gliders were made available.[17] The gliders, to be used for ferrying guns and heavy equipment, would land during the intervals between the drop of parachute echelons. Bad weather prevented or delayed the arrival of many of the transport aircraft. Fewer aircraft for sorties meant carefully worked out loading plans by airborne units had to be readjusted. The initially precise planning would now have to be redrafted while other operational considerations competed for the planners' limited processing capacity and precious time. To make matters worse, pilots refused to carry full loads in turbulent winds caused by the passing weather front. Transport pilots reduced loads from twenty paratroopers (or drop units) to only fifteen or eighteen per aircraft.[18] Faced with these last minute limitations, many commanders chose to eliminate supplies and heavy weapons to load more men for the initial jump into the drop zone.

Delayed confirmation for the mission and time consuming coordination between Voronezh Front and the airborne corps left little time for lower unit leaders to make plans and adequately brief their inexperienced troops. Battalion commanders conducted their operational briefings on the runways. Company commanders who learned of battle positions immediately before loading onto waiting aircraft had little or no opportunity to review contingency plans in case of wide dispersals, immediate contact with the enemy, or the unexpected. The Soviets needed a perfect drop.

The 3rd GAB planned to land in the area of Medvedovka, Novoselki, and high ground two kilometers southeast of Novoselki. The brigade was to be dropped in two waves: 1st, 3rd Battalions and then 2nd, 4th Battalions.[19] The 1st Battalion had the mission of taking the towns of Lopovy Rog and

Pii. The battalion was to set up a defensive line to prevent the approach of German units from the west into these towns. The battalion's right flank was oriented for link-up with friendly forces, and reconnaissance troops were to establish contact with elements of the Red Army at Shchuchinka.[20]

South of 1st Battalion, also facing west was the 3rd Battalion in Novoselki and Makedony. The 2nd Battalion was to dig in along a line from Schandra to the lake at Sinyavka. The 4th Battalion was to occupy a second echelon defensive position at Gruschevo, Kalinovka, and Potaptsy behind the three forward battalions. The signal to mark assembly areas was red and white flares. The haste of the 2nd Battalion commander's briefing was evident in his concluding notes' illegibility to German intelligence personnel who read his captured notebook.[21]

The 5th GAB was to land west of Kanev to block routes from the west and southwest preventing an advance of German forces into the area. The brigade was to establish its battalions along the Rossava river protecting the southern flank of the 3rd GAB at Sinyavka. The 2nd Battalion was to be in the center sector with 4th Battalion on the right and 1st Battalion on the left.[22] Once on the ground the plan required battalions to collect at battalion assembly points. Troops would gather on various colored pyrotechnics. The 2nd Battalion, for example, would be identified in the dark by green flares or green flashlights.

The 3rd GAB commander, Colonel V.K. Goncharov, loaded his entire brigade staff element on his plane to use flight time for last minute plans and decisions. However, he did not load any radios or radio operators on the plane.[23] The first aircraft for the 3rd GAB took off around 1830 hours.

The 5th GAB's departure from the Bogodukhov airfield was not without incident. Although the brigade did not experience the same last minute scramble for mission planning as the 3rd GAB, the 5th was forced to make last minute improvisations in loading plans. The brigade staff expected to receive sixty-five transports in the first wave, and plans were minutely tailored to this airlift capability. But only forty-eight aircraft arrived. The situation deteriorated when four of the gasoline pumps for refueling aircraft malfunctioned. Zhukov's promised intervention with the arrival of fuel trucks did not guarantee that they would work.[24] The time allocated to refuel all planes was too short. Some previously loaded paratroopers had to off-load and march across the tarmac and reload on fueled aircraft. Reloading delayed take-off time for the first flight by one and a half hours, and ten planes were bumped into the second flight.[25] The 5th GAB, loaded in two lifts, used all available aircraft. The first lift, comprised of the 1st and 4th Battalions, took-off around 2030 hours. The second lift, 2nd and 3rd Battalions, departed at approximately 0200 hours.

At the third airfield, Smorodina, where the 45mm field guns had been assembled for the three brigades, not a single aircraft could take off that night. The same problems of coordinating proper refueling trucks, fuel, and lubricants broke down. Consequently, Red Army paratroopers would fight with no organic artillery fire support.

On the evening of 24 September, a total of 296 sorties instead of the planned 500 were made. Approximately 4,575 paratroopers and 666 cargo containers were dropped. A total of 2,017 men and 590 cargo containers and all 45mm field guns failed to reach their drop zones.[26] From the airfields the distance to the drop zones was 175 to 220 kilometers. The flight time required about one hour and twenty minutes from Lebedin, two hours from Bogodukhov.

Unsuspecting paratroopers, whose command believed that they would be dropped into relatively clear drop zones, faced a situation far less certain than ordinarily required for a successful operation. Bad weather precluded the dropping of parachute scouts to obtain reliable, current information on the enemy in the drop zones, the establishment of contact with partisans, or marking drop zones with lights and fires. Despite the lack of these critical preliminaries, the planning momentum continued to push the operation forward.

Inexperience apparently kept alive the belief that the operation would achieve its objectives. Inexperience also failed to see that delays had pushed the operation beyond the optimum time for reasonable success. No one was willing to revoke the promise - however elusive - that a rapid strike behind German lines seemed to offer.

The large airborne drop required great courage and skill not only of paratroopers but also of transport pilots. Flying planes for an airborne drop is technically demanding and operationally hazardous. Pilots must fly in close formation at low altitude at a relatively slow speed of 120 miles per hour. But, in this operation these challenges were exchanged for others - for refueling delays and the pilots' inexperience in formation flying insured that transports were strung out as they arrived over the front lines.

When the aircraft approached the front lines, German air defense units opened fire with 20mm antiaircraft guns. Flying through flak is crucial for a successful airborne drop. A number of Soviet pilots, however, seem to have panicked, become disoriented, and signalled airborne units to jump. Paratroopers stepped into the dark too early. Sticks of paratroopers landed on the east side of the river among Red Army units, while others dropped into the Dnepr. Not all pilots followed through with the drop; thirteen planes returned to their airfields with jumpers still on board. German reports indicated three aircraft and one glider were actually downed. This low count testifies to the Soviet pilots' overreaction to antiaircraft fire and their failure to fly through to drop zones.

The Red Air Force transport, Li-2, had an exit door 70 centimeters wide. It was impossible for paratroopers to plunge through the exit rapidly with large weapons or equipment. Each paratrooper jumped with a machine pistol, sub-machinegun, or rifle, two grenades, 200 rounds of ammunition, three days' rations, and a small shovel. They were not issued ponchos despite the slight autumn frost at night. For the necessary quick, steady flow out of aircraft sides, the parachutists carried very little.

However, to compensate for lack of equipment on individual jumpers, cargo containers were used to carry extra arms and equipment. These containers weighed an average 175 pounds. Two to four of these containers per aircraft had to be dragged to the doors by hand and pushed out of the exit. Understandably, all this took time, slowing parachutists' exit time, spreading units across the ground.

In many cases parachutists were forced to jump while their aircraft was taking evasive action against antiaircraft fire. Although paratroopers trained to jump at altitudes between 400 and 600 meters, they exited aircraft at altitudes ranging from 600 to 2000 meters and some at speeds of 200 kph. When they exited planes, some jumpers noticed strong winds that would further spread their units.[27]

Mixing of paratroopers before the take-off, uncontrolled dropping, and failure to drop a large number of the force ended any rational hope for consolidation on the ground by units. On this point, unfortunately, Soviet sources are mute. The impression is that many of the jumpers, once they hit the ground, could not find their assembly areas. Others searched frantically for the cargo containers with heavy weapons. A few used their illumination flares to find their cargo containers. Five jumpers of the 5th GAB could find neither their assembly area nor their cargo containers with their heavy weapons. After a slight resistance, they surrendered.[28] These green troops on their first combat jump at night were lost, lonely, and afraid. They were left in groups too small to be effective.

The 3rd GAB was dropped 20 to 25 kilometers short of its intended drop zone. The greater part of both brigade waves landed in an enormous area bounded by Dubari, Troshchin, and Kolesishche. The 5th GAB was unintentionally split between two drop areas. The first wave with the brigade staff was dropped generally in the northern area. The second wave, or at least parts of it, as indicated by captured documents and maps, landed in the vicinity of Yablonov.

Elements of both the 3rd and 5th GABs dropped in the 112th Infantry Division sector the first night. The 112th reported continuous landings throughout the night in the areas of Kovali, Borbritsa, and Kolesishche, and the division became engaged in hard fighting with large groups of Red Army jumpers.[29] The division waited for dawn before it committed units against the parachutists.

Other German units along the river reported paratroop drops under cover of darkness between 2000 and 2200 hours, and immediately engaged them. The 19th PD, farther west, reported a quiet night, although it was aware of paratroop landings in the areas of Kolesishche, Dubari, and south. Only a small portion of the division was engaged in fighting paratroopers at Dubari.

During the morning of 25 September, German air reconnaissance observed 1400 parachutes on the ground generally along a line west from the actual drop areas because of strong winds. Although Red Army paratroopers had been taught to conceal their parachutes, jumpers gave no consideration to cargo chutes.

The 3rd GAB commander rallied about 150 paratroopers on the first night. They moved west to a small wooded area east of Grusheva. The problem for most of the 3rd and part of the 5th GABs in the northern region was lack of forested areas for hiding. The terrain was sparsely wooded and cut with ravines and streams that afforded little or no concealment. The movement of paratroopers resembled that of mice in a maze. Men scrambled through streams and ravines which conformed to no sensible pattern. The open high ground had to be avoided. With daylight, German reaction forces could respond to the reports of scattered paratroopers and easily track them down. Villagers found groups of paratroopers, many already casualties from firefights with German soldiers, huddled in deep ravines.[30]

On 25 September, instead of the ordered attack against Red Army forces at Grigorovka, the 112th Infantry Division fought paratroop groups and reported killing 300 paratroopers and capturing 67 prisoners, including seven officers of the 5th GAB staff. Despite reported heavy losses to the Soviet landing force, the 112th believed only a small portion of the paratroopers had been cut down. The number of paratroopers raised concern both for the division's lines of communications and for the possibility of being cut off in the face of heavy pressure from Red Army ground units in the bridgehead.

The 19th PD was primarily concerned with ground attacks south from the bridgehead. The division dispatched part of an infantry battalion to deal with the 150 paratroopers under the 3rd GAB commander east of Grusheva. After a very stiff fight, the commander was captured. The 19th also reported a downed "Douglas" transport and one heavy glider, each with twenty dead.[31] The paratroop-laden glider, which was to carry supplies and heavy weapons, indicates that some Soviet commanders opted for additional men in the initial assault and were willing to wait for subsequent supply drops after securing their objectives. It also indicates that the gliders showed up at the wrong airfields, instead of Smorodina where they were to lift the 45mm guns.

The 10th PGD reported destroying a group of twenty paratroopers in a small wooded area one kilometer east of Vesselaya Dubrova (SW of Shchuchinka). The drift of larger paratroop units into their sector would come over the next few days.

The 57th Infantry Division reported it had been fighting with scattered paratroop groups since early morning 25 September. The majority of these jumpers were in the woods east and north of Yablonov and northeast of Litvinets. The division reported the capture of fifty-six prisoners, nineteen enemy dead, and captured "booty," ammunition, weapons, and explosives.

By the end of the first day, XXIV Pz Corps' assessment of the situation was based on information from prisoners and captured documents. With these sources, German intelligence identified the airborne missions and composition. The German leadership also knew that the 1st GAB was poised for a drop on the night of 25/26 September. Accordingly, German units alerted their rear area troops.

As a result of hasty preparation and loading, serious oversights became obvious. The 3rd GAB commander was not the only commander to overlook the loading of radios and radio operators. These critical communication links became lost in the confusion. Many radio operators were separated from their radios during the drop. In confusion on the ground, other operators concerned about the capture of their radios buried them for future retrieval. Without radios, airborne brigade commanders could not communicate among themselves nor with the airborne corps staff and other Red Army units.

Lack of radio contact with the landing force prompted an important decision by a Soviet High Command ill-disposed to reinforce failure. With no idea of the fate of dropped forces, Joseph Stalin, who was closely following the operation, called a halt. The 1st GAB and other remaining assets were not to be dropped. Stalin had been annoyed by misuse of the airborne forces and criticized the night drop decision. He decided to return remaining assets to Stavka reserve.[32] The 1st GAB was never dropped at Bukrin.[33]

Although no significant hostilities occurred between paratroopers and German divisions during the night of the 25th, on the 26th, the 57th Infantry Division reported several morning attacks just north of Pekari. Small Red Army units landing from the east bank were joined by paratroopers who came from the northwest. Another surprise attack with nearly 200 on Pekari from the southeast occurred late morning. The Soviets temporarily overran part of the division. Despite heavy losses requiring the commitment of other divisional elements, the 57th regained the area by evening. Soviet soldiers fought tenaciously to defend themselves, but were pushed south of Pekari against the Dnepr.

Farther north, the few wooded areas that promised cover and concealment were lodestones, drawing exposed paratroopers to their destruction. On 26 September, the 10th PGD encountered a group of 150 to 200 paratroopers east of Vesselaya Dubrova, in the same area where they had destroyed twenty paratroopers the day before. The Germans inflicted heavy casualties on the jumpers and forced them south away from the Red Army units in Shuchuchinka.

While fighting a large group of paratroopers in a ravine south of Buchak on the boundary with 57th Infantry Division, the 112th Infantry Division repelled a number of Red Army attacks from the bridgehead south of Grigorovka. Between 0000 hours on 24 September and 1600 hours on the 25th, the division had lost 200 soldiers and was expecting Red Army ground forces to be strengthened around Grigorovka.

During the night of 26 September, 19th PD reported equipment destroyed by marauding paratroopers. The 10th PGD also experienced isolated groups of paratroopers drifting through its sector attempting to link up with Soviet units at Shchuchinka. Rear area forces continued to guard to the south and southeast of the division against a number of paratroopers who were apparently driven into the 10th PGD sector by 19th PD.

On 27 September, new paratroop groups appeared west of Khodorov and southeast of Shchuchinka as the 3rd GAB moved from the northern area to link up with units of the 40th Army. These remnants engaged parts of three German divisions. Without accomplishing their mission, individuals and small groups slipped to safety. If any escaped to Red Army units, their fortune has not been recorded. German rear support troops and designated security forces from XXIV Pz Corps units systematically destroyed numerous paratroopers, reducing the airborne threat to a level which merited no mention in German division daily journals after 27 September.

In contrast with the experience of XXIV Pz Corps after 27 September, III Pz Corps continued to make contact with paratroopers well into October. Large and numerous forested areas in the region southwest of Kanev provided a different environment for groups of the 5th GAB and the few 3rd GAB parachutists who drifted south. Lieutenant Colonel Prokopii Sidorchuk, commander of the 5th GAB, who had landed in the vicinity of the Kanev forest, steadily assembled paratroopers. He continued his activity into the first week of October, raising his force to 600 men.[34] During October, this growing group of paratroopers moved among major forested areas behind German lines, establishing contact with partisans.

During the period 28 October to 11 November, reconnaissance and sabotage groups of the brigade blew up bridges, attacked supply convoys, and raided German garrisons. Soviet military writings on the Bukrin drop have focused on this, the one positive aspect of the operation.

The 5th GAB remnants eventually established communications with the 2nd Ukrainian Front and maintained direct contact with headquarters, 52nd Army. On the night of 12/13 November the 5th GAB provided west bank support to the crossing of the Dnepr river by 52nd Army. The Red Army finally pushed their last elements across the Dnepr in the area they had first hoped to cross. Lieutenant Colonel Sidorchuk brought out perhaps 1000 paratroopers, the survivors of the drop which had occurred nearly two months earlier. Their ordeal was over.

Although the Red Army showed initiative and a willingness to accept risk, their largest airborne drop in World War II for the Burkin bend was an opportunity missed and an operational disaster. In warfare, as in most human endeavors, initiative is the ability to spot and exploit a fleeting opportunity. The Red Army leadership clearly saw the opportunity to preempt the German defense along the Dnepr river. As the Red Army moved to implement their plan for crossing the river with the use of airborne forces, general inexperience exacted steep costs. The Soviets failed to master the subtle techniques and mechanics of execution that are so crucial in airborne operations. Paratroopers and pilots were inadequately trained, limiting the skills to be applied creatively. The air force fell short of the required number of transports, and plans were not designed with enough flexibility for contingencies necessary in an airborne operation. Coordination failed as staffs miscalculated timetables for rail movements and fuel trucks. All these factors began to limit the initiative and creativity of the operation. An experienced understanding of one's forces, their equipment, and the capabilities of the personnel, is a prerequisite for creativity. Without this technical mastery, the "friction" of inexperience shows. So elusive is opportunity that it often disappears before inexperience.

NOTES

1. The operation has either been ignored or given only brief acknowledgement in official Soviet histories such as the six volume *The Great Patriotic War, The Short History of the Great Patriotic War* (interestingly, the Russian edition has one paragraph while that paragraph is omitted in the English edition), and the recent 12 volume, *History of the Second World War.* For specific works on Soviet airborne see G.I. Sofronov, *Vozdushnye desanty vo vtoroi mirovoi voine* (Moscow, Voenizdat, 1962); I.I. Lisov, *Desantniki* (Moscow, Voenizdat, 1968); and D.S. Sukhorukov et al., *Sovetskie vozdushno-desantnye* (Moscow, Voenizdat, 1980). Significant articles are Ya. Samoilenko, "Iz opyta upravleniya vozdushnymi desantami v gody voiny," *Voenno istoricheskii zhurnal* (hereafter *Vizh*), 12 (1979), pp. 15-21, and D. Sukhorukov, "Byvody iz opyta primeneniya vozdushnikh desantov vo vtoroi mirovoi voine," *Vizh*, 7 (1981), pp. 65-74.

2. Walter K. Nehring (commanding general XXIV Panzer Corps in the fall of 1943), "Employment of Russian Parachute Forces in Cherkassy-Kiev Area (23-24 September 1943)" and Eitel-Friedrich Binder (operations officer 19th Panzer Division), "Employment of a Russian Parachute Brigade in Bend of Dnieper Northwest of Kanev," *Russian Airborne Operations*, MS #P-116 (Department of the Army, Office of the Chief of Military History, Washington D.C., 1952); *Airborne Operations: A German Appraisal*, Department of the Army Pamphlet 20-232 (Oct. 1951), pp. 36-37; Otto von Knobelsdorff, *Geschichte der neidersachsischen 19. Panzer Division* (Bad Nauheim: Verlag Hans-Henning Podzun, 1958), pp. 228-232; Paul Carrell, *Scorched Earth* (New York, Ballantine Books, 1966), pp. 401-411.

3. Erich von Manstein, *Lost Victories* (Presidio Press, 1982), p. 458.

4. Lisov, p. 155. Lisov claims that he was chief of staff of the control group for this large exercise. However, interrogation reports and German intelligence summaries do not indicate any major unit exercise for those captured at Bukrin.

5. Ibid.

6. XXIV Panzer Corps, Ic Tatigkeitsbericht (Intelligence Reports), "Evaluation of Captured Documents and Prisoner Statements of 3rd and 5th Guards Airborne Brigades," undated, p. 1 U.S. National Archives Record Group (RG) T-314, roll 2318.

7. Ibid.; III Panzer Corps, Intelligence Reports, "Prisoner of War Interrogation Report," (28 September 1943), p. 1, RG T-314, roll 205.
8. 11th Panzer Division, Intelligence Records, "Prisoner of War Report" (29 September 1943), p. 1, RG T-315, roll 603.
9. Nikolai S. Skripko, *Potselyam blizhnim i dal'nim* (Moscow, Voenizdat, 1981), p. 323.
10. Ibid.
11. A.M. Zvartseva, *3-ya Gvardeiskaya tankovaya* (Moscow, Voenizdat, 1982), p. 94.
12. Nehring, p. 40.
13. XXIV Pz Corps, Intelligence Records, p. 1.
14. 11th Panzer Division, p. 1.
15. Sukhorovkov et al, p. 197.
16. Sofronov, p. 30.
17. The Li-2 was the Soviet license-built version of the Douglas DC-3. The C-47 was a U.S. military version of the DC-3. Under the U.S. Lend-Lease to the Soviet Union, 707 C-47s were delivered to the Soviet Union over the Alaska-Siberian Air Ferry route between June 1941 and September 1945. How many of the Li-2 were in fact Lend-Lease C-47s for the Bukrin operation is not known in available sources, but some were undoubtedly present.
18. Sofronov, p. 30.
19. XXIV Pz Corps, Intelligence Records, p 2.
20. Ibid.
21. Ibid.
22. III Pz Corps, 2.
23. Sofronov, p. 32.
24. Skripko, pp. 323-324.
25. Sofronov, p. 31.
26. Lisov, p. 163.
27. 11th Panzer Division, Intelligence Records, "Prisoner of War Report" (27 September 1943), RG T-315, roll 603.
28. III Pz Corps, p. 2.
29. XXIV Pz Corps, Ia Kriegstagebuch (Operational War Journal), "Daily Report" (25 September 1943), p. 3, RG T-314, roll 724.
30. I.I. Lisov, *Zemlya-nebo-zemlya* (Moscow, Izdatelstvo DOSAF, 1973), p. 68.
31. XXIV Panzer Corps, Operational Journals, p. 2.
32. S.M. Shtemenko, *The Soviet General Staff at War, 1941-1945* (Moscow, Progress Publishers, 1975), p. 184.
33. One unusual development was the discovery of the body of an officer at Dubari bearing 1st GAB documents. The corpse was probably a liaison officer from the 1st

GAB who jumped with the 3rd GAB to facilitate coordination for subsequent drops by 1st GAB. Some confusion in the reporting of this finding led German Eighth Army to believe that all three brigades had been dropped on the 24th. The information was inaccurate, yet many current western accounts erroneously identify the 1st GAB as a participant in the drop.
34. Lisov, Desantniki, p. 167.

AUTHOR NOTE: Translation of German records by Trudie Thompson and Thomas Rauter.

CHAPTER THREE

Red Army Tank Corps Commander

Rarely do we possess the necessary insight for an accurate assessment of new military phenomena. Understanding comes with time, and with time we gain experience. Primarily through experience, we develop a character of thought that is flexible and imaginative, which allows us to make proper and incisive assessments. One method of expanding individual experience is to study the past. Breaking the bounds of one's immediate environment and time-span allows one to see a richer variety of alternatives and solutions to current problems.

Currently, there is a new military phenomena that has evoked great speculation.[1] For the modern military professional, many unanswered questions are raised by the sudden awareness of the Soviet Operational Maneuver Groups (OMG). How will the OMG be committed? What will be its command and control? How will its operation be supported in the depths of a penetration? All these questions are legitimate concerns for an understanding of how to fight the OMG.

Yet, the dry, clinical speculations and facts regarding OMG operations give no animation to an intuitive understanding of its employment. If one could take part in an OMG attack into the operational depth of an enemy defense, one's assessment would be richly enhanced by such an experience. While accompanying an OMG operation today is impossible, one can gain some useful insights from a knowledge of Red Army World War II mobile group operations. Soviet mobile group operations serve as the fundamental combat experience for Soviet Army military theorists and OMG planners, who strongly insist that the study of World War II experience is a prerequisite to understanding contemporary combat.[2] In essence, the missions and functions of the two groups are the same since the OMG use is strongly influenced by the mobile group experience, and primarily the impact of technology alters techniques and capabilities in combat.

Through a composite of mobile group operations, we may very well create an appreciation for standard Soviet organization and routine procedure in armored operations of this nature. This background will develop a useful historical frame of reference for understanding OMG potentialities today.

Since most mobile groups were based on tank, mechanized, or cavalry formations, we will use for our purposes the actions of a typical Red Army tank corps in 1944. By 1944, Red Army tank units were hitting their stride in force structure and competency of commanders. A tank corps at that time consisted of one motorized rifle brigade and three tank brigades. It had a number of combat support units. On average, with 12,000 men and 200 tanks, 63 self-propelled guns, 182 artillery guns, mortars, and rocket launchers, the tank corps packed considerable fighting punch for its day.

Red Army Pre-Planning for Operations

We will accompany the tank corps commander through a typical breakthrough operation. Although he is relatively young, the commander is experienced and holds the rank of major general. He started the war in the summer and fall of 1941, as a tank brigade commander; in the spring of 1942, he became a tank corps commander. He has been fighting the Germans for three years as a tank commander and is very professional at his job.

As commander of the 13th Tank Corps, he is informed by courier that in the approaching operation his corps will be committed as an army-level mobile group into a breakthrough made by the 30th Guards Rifle Army. His mission will be to exploit tactical success by attacking into the German defense's operational depth. Since commitment of the tank corps to action will be one of the decisive moments of the attack, an army-level staff officer is sent to the corps command post to assist in planning and to assure close, continuous liaison and coordination.

The corps commander, his brigade commanders and key staff section chiefs take a trip into the area where the breach will be entered. They proceed to reconnoiter the intermediate assembly area and the routes leading to it. They also get a general orientation of the situation from commanders of forward rifle units in contact, along whose sectors the breach will be entered. Clad in ordinary soldiers' uniforms with no distinguishing ranks, the corps commander and his senior officers visit the front for a first-hand look at the terrain.

Knowing the general situation, the corps commander begins to prepare his tank corps units in a concentration area approximately 75 kilometers from the front line. Since the corps can be located in the concentration area from three-four to fifteen-twenty days, he organizes staff exercises for his staff and subordinate staffs. The commander, using a terrain mock-up, takes the subordinate commanders through the operation explaining alternative actions at key points in the operation. These pre-operation exercises ensured that subordinate commanders understood the corps mission and provided for a concentric effort by subordinate units in the confusion of combat.

One of the primary operations to be planned and coordinated is the actions of the forward detachment. A forward detachment was usually a reinforced tank brigade that had the specific mission of seizing and holding important tactical or operational objectives. It had the additional missions of disrupting German defenses in depth and partially preventing the maneuver of the enemy's operational reserves. commanders of forward detachments were selected from the decisive and most experienced officers.[3] It was not unusual for the army or front commander to visit and ask for detailed information about the preparation of these detachments. They talked with the forward detachment commanders and gave them specific instructions.

The success of a forward detachment's operation was directly influenced by establishing an optimal composition that allowed it to operate with autonomy. A tank or mechanized brigade sometimes included up to a regiment of self-propelled artillery and a battalion of antitank and antiaircraft artillery, and up to a battalion each of rocket launchers, combat engineers, and motorized infantry. Dependable radio communications were established. The extensive employment of forward detachments in offensive operations became an integral part of Red Army armored forces in the operational depth because of the number of diverse and complex combat missions they could perform.[4]

The corps commander and responsible staff officers met personally with brigade and battalion commanders. Before World War II, missions to subordinates were passed through written operations orders. As Red Army leadership gained experience, they realized timeliness was of the greatest importance, and transmitted key orders orally in meetings between commander and subordinates. Written confirmation followed later. Many Red Army commanders believed no written order or telephonic directive replaced giving orders personally, on the spot. The commander has the opportunity to make certain that instructions are correctly understood by subordinates, and to resolve all unclear questions - not only on a map, but also in the field.

The political affairs cadre conducts classes in preparation for the coming fight. Often, as the corps moves closer to combat, these sessions by political officers are conversations with anxious men to boost morale. In the mind of Red Army commanders, indifference is a disease that blunts vigilance and is prevented at all costs.[5]

In preparing his staff, the tank corps commander determines specific actions to be taken by all the officers of his staff during preparation for the penetration and during subsequent battles. This avoids assigning additional missions for tactical control during battle.

At this time the commander determines the make-up of his operational group. The operational group is a mobile command post which moves with the commander. Small in size, the operational group usually has the commander, the corps artillery commander, chiefs of operations and intelligence staffs, liaison officers with higher and subordinate commands, and a signal communications elements.

Movement to Battle

From the concentration area, the 13th Tank Corps moves forward to a designated intermediate area. Reconnaissance of the intermediate area is conducted. The units move at night and, as they arrive, they are hidden in the forested areas. Traces of tank tracks along the various routes into and in the intermediate area are obliterated. A ban on all radio communications prior to commitment in the operation is observed strictly. Patrols search the area to clear any possible enemy reconnaissance. The army or front has created a 25-kilometer exclusion zone behind the front line from which the local population is evacuated.

Often the immediate deployment of the tank corps is hidden not only from the enemy but also from friendly front line troops. Only the forward commander and a small portion of his staff for coordination of the passage of lines knows the true mission and location of the tank corps.

Intermediate areas, situated to move the tank corps in the direction of the impending operation, provide concealment to the closest possible distance from the line of commitment. Here final preparations are made. If only 15-20 kilometers from the line of commitment, the tank corps remains a short time, topping off vehicles with fuel and making last minute maintenance repairs. An intermediate area this close to the front line precludes designating an additional jumping-off position for these final preparations. The tank corps moves directly into the attack.

Just prior to commitment, the corps commander locates with the 30th

GA commander's forward command post, which allows the army commander to ensure the organized commitment of the tank corps into battle. From this point, the corps commander watches the timing of his corps' advance in order not to arrive at the breach too early or too late. Too early risks unnecessary casualties, and too late risks a closed penetration or responding enemy reserves at the line of commitment. Upon direction from the Army commander, the corps commander moves to the observation post of the forward rifle corps commander where the commitment of his corps will take place.

During operations, operational groups headed by the corps commander are often established to function as a forward command post. Red Army commanders wanted to be well forward and at decisive points in the battle.

Coordination continues with the front line unit in whose zone the tank corps will be committed. The forward unit clears the roads; all supply and transport vehicles are routed off the main roads. Priority goes to the 13th Tank Corps.

Sometimes, at the last moment, the situation could dictate a redirection of the tank corps to another sector, in which case the coordination has to begin anew and hastily. The corps commander had heard of General D. D. Lelyushenko, commander of the 4th Tank Army, discussing his major redirection during the Lvov-Sandomierz operation:

At 1400 hours on 16 July 1944, General Lelyushenko received from the Front commander orders to leave two of his brigades in the 38th Army's sector and move north to 60th Army's sector to be committed at dawn on the 17th behind the 3rd Guards Tank Army in a 4-6 kilometer wide penetration. All the questions of coordination and support had to be done over again in a matter of hours.[6] The operation was successfully organized and executed on the move.

The 13th Tank Corps chief of operations directs work on combat orders and coordinates with forward units of the 30th GA. The chief of intelligence collects and processes information from higher and forward units. Red Army tank commanders like accuracy and do not tolerate "approximations" in estimates by their intelligence officers. For tank forces moving into the operational depth, the location of enemy reserves is most important.

As the corps moves towards the line of commitment, the brigades and supporting units are arranged to preclude regrouping and loss of time. In this operation, because the approaches to the front line are exposed to enemy aerial and ground observation, the corps advances quickly and directly into battle.

The Chief of Technical Services advises the corps commander that tanks which had broken down enroute from the concentration area to the intermediate area are repaired, and the units up to strength in tanks. The front line army, 30th GA, will provide maintenance support until the corps commits through the breach, leaving the corps maintenance support intact for support behind the enemy lines.

While the tactical configuration of the tank corps depends on many variables such as combat mission, fighting strength, width of sector, enemy defense, and terrain, the tank corps commander usually seeks to provide for flexible control and speed. Based on the information concerning the situation, the commander prescribes a zone six to eight kilometers wide in the advance to the line of commitment. This gives him room to maneuver the corps. He forms the corps into two echelons, an artillery group, and a reserve. Two tank brigades, with a large portion of the support weapons and all the artillery, formed the first echelon, while advancing the corps on two axes. The depth of the columns of the main forces reached 20-22 kilometers (see figure 2).

Questions of coordination and passage of lines with the commanders of the first echelon rifle corps were taken up previously on the day preceding transition to the offensive. At that time, the tank brigade commanders conducted a reconnaissance of routes of advance.

An attached representative from the supporting air army worked out aviation support. In coordination with the operation, the air representative planned air support for the approach and supporting strikes once the unit broke into the operational depth.

For greater flexibility and agility in signals communications, selected combat vehicles with cross-country capabilities were fitted out as main radio stations and the commanders and staff ran a communications test before moving out. Communications within the corps are organized in the following manner: corps staff has radio communications with the subordinate units on two channels; with staffs of higher headquarters over three channels; and two stations for communications nets to other units and commanders.

The tank corps commander from his operational group has a radio station working with the brigade commanders and a radio station with the higher commander. Also, on the corps commander's net with the brigade commanders is the chief of staff for the corps, which allows him to keep informed on orders and reports from brigade commanders.

The radio nets for artillery are organized in a similar manner. The corps staff maintains separate channels with reconnaissance and the rear, and a separate net with the front line infantry who will create a breach in the line for the tank corps. Through radio nets, officers from the air units assigned to the corps staff communicate with air support to direct aircraft to targets.

Under the cover of darkness, the reconnaissance elements of the first echelon brigades begin the advance on their respective axes. Reconnaissance parties consist of a platoon of tanks, a section of combat engineers, up to a platoon of submachine gunners, and armored vehicles and motorcycles.

The Movement Support Detachment follows behind the reconnaissance and begins necessary obstacle clearing on the movement routes for their parent brigades. Approaching the line of commitment, forward detachments deploy into combat formations. The corps commander receives the signal from the rifle army commander to commit to battle.

The Battle

The corps commander hopes to have a "chistaya" (clean) breakthrough. This means the first echelon rifle army has already penetrated the entire German tactical defense zone, clearing a path into the rear area. If this is the case, the tank corps commander can leave the rest of the corps in column for speed and ease of movement.

Most of the time, as in this case, the infantry fails to breach completely all the tactical defenses. Adding its tank strength, the tank corps also fights to complete the breakthrough before moving into the operational depth.

The commander orders the forward detachment to assist the infantry in creating a breach. The breach, as the zone of commitment for the tank corps, is too narrow with only a single road for commitment. The commander knows from experience that German commanders will bring up local reserves with the mission to restore the situation and close the "throat" of the penetration. In the interest of seizing this fleeting opportunity and maintaining the tempo of the attack, he risks commitment in the narrow penetration.

At least during the initial contact with counterattacking enemy reserves, the tank corps will have the rifle army's artillery and close air support. This will give the tank corps a local numerical advantage that would preserve his force and allow him to move into the depth of the defense.

The tank corps will ignore flank fires and will avoid engaging attacks from the flank in order to speed through the gap. Securing the shoulders of the penetration is the concern of the first echelon rifle army, its reserves, and any available reserves from Front level.

In this case, prior planning for tactical control at this point is effective.

The corps commander with his operational group moves directly behind the combat formations of the first echelon of the left column. He controls his forces by issuing short combat orders by radio; liaison officers carry duplicate messages. Radios transmit only after the corps and brigades begin moving up to the attack line. At this point, the pace of the action at the tactical level is too fast for reaction by German forces to any signals interception. Army staffs had previously worked out a unified coded map and code table.[7]

The main command post with the corps staff leap-frogs forward. The rear command post to the main, main command post to the forward. The movement of the command posts depended on the rate of advance. By the middle of 1944, Red Army High Command promulgated directives, based on analysis of combat experience, and established the practice of working out of a specific plan for the movement of basic command posts. Headed by the deputy chief of staff, an officer group with a communication group moved ahead of time to the new location.

The command posts of armored formations displaced in large jumps of 15-30 kilometers with the existing communications means restricting the number of command post moves to not more than two or three a day. The corps commander's operational group was more mobile because its move did not involve the amount of time in changing location as the main command post. In some operations, the operational group halted for twenty-thirty minutes every hour and a half of movement.[8]

The speed of deploying and taking down command posts depended largely upon the teamwork of staffs and prompt preparation of the equipment and communications. The practice of setting up command posts according to a previously elaborated scheme proved effective. In this instance, fifteen-thirty minutes were spent in a brigade, and forty-sixty minutes in a corps.[9]

In the Operational Depth

As the operation unfolds, German units in sector have four previously-prepared defensive lines. The infantry has broken through the first two defensive lines, and the remainder require breaching by the tank corps. Rushing past infantry units, the tank corps punches through the German screening forces and reaches the third defensive line. The advance holds up here. Being well forward, the corps commander assesses the situation in a timely manner and makes the decision for the execution of a turning maneuver. He covers his right flank with a tank unit of the second echelon and assigns the rest of the second echelon the mission of attacking the Germans from behind the left flank of the first echelon. He brings up artillery. At the same time, he directs the fire support for the assault of the first echelon. With the movement of the second echelon around the left flank, the Red tankers threaten German defenders with encirclement and an attack from the rear. Shifting to meet this threat from the rear, the German commander weakens his front. Taking advantage of this, the tank corps' first echelon breaks through the German resistance and scatters it.

Once committed through the tactical defensive zone and into the operational rear, the tank corps commander receives direct support from air units subordinated to his command. Air strikes clear his way. The tank corps' momentum carries it through the fourth line of defense. The German fire plan has gaps, and they have not had time to emplace obstacles and other engineer barriers to slow the tank attack.

The corps commander watches his casualty rate and tank losses closely. He has lost close to twenty tanks. This is ten percent, and experience has taught the Red Army that they could tolerate up to twenty-five percent losses in these initial phases and remain combat effective. His corps is in good shape.

As his tank corps emerges through the German tactical defense zone, the commander worries about other concerns. One of the important requirements for mobile groups at this stage of the operation is to achieve a high rate of advance. At the operational level of war, high tempo begins to disrupt the whole orderly commitment of the enemy reserves to action, and it is this breakdown of the defense's systematic control and unit integrity that provides the basis for successfully developing an operation to a great depth.[10]

The commander also knows that as the tempo of his forces picks up, the corps at this time has become more elongated. German units will undoubtedly seize roads in the rear. The commander is concerned that his staff is not cut off. He directs their movement to a position close behind the main forces of the corps.

The development of the offensive tempo depends greatly on how quickly mobile forces cross water obstacles. Quite often, throughout the war, rapidly moving tank forces in the operational depth were forced to cross several obstacles during one operation. The chief method was to cross on the move. The forward detachment would seize crossings and bridgeheads. All possible measures were taken to ensure artillery and bridging materiel moved up to the river quickly. Supporting aviation would be tasked to isolate crossing sections from any approaching enemy reserves.

The commander advises the Chief of Technical Services to ensure the organization of maintenance reconnaissance groups to quickly identify repairable tanks and get them back into service. Tanks and assault guns coming out of repair were a main source for replenishing tank units in the course of an operation.[11] Red Army experience in the latter period of the war revealed that some tanks were "recycled" three or four times during an operation.

Other matters of supply do not bother him at this time. His corps had begun the operation with three and a half times the combat load of ammunition, four times its fuel and lubricants, and they carried almost fifteen days of rations. And, they are prepared to forage for immediate needs. At this stage, units are expending more fuel than ammunition. German fuel dumps would be used for refueling before being destroyed.

Having beaten the initial reserves of the German units, the corps commander knew that for the next several days, they would not meet any serious resistance. Then, they would run into reserves from the strategic depth. In the meantime, the tank corps actions in the depth of the German defense would be demanding. The continued spread of units on separate axes would make tactical control difficult. Brigade commanders operated more independently, displayed initiative and acted aggressively, to break up and cut off withdrawing Germans forces. At this point in the operation, the corps commander has to affect very careful coordination and assignment of missions to his brigades. He and his deputy would go to the brigades personally and assign missions on the spot.

The chief of intelligence provided very important information. To effectively operate in the operational depth it was necessary to determine in advance the routes of withdrawing German forces and the strength of garrisoned strongpoints. Most of the time, it was necessary to bypass these centers of resistance. Mobile groups avoided any battles of attrition. Early warning of the advance of enemy reserves was of special importance.

After several days operation, the tank corps had lost more tanks to battle and mechanical failure, and supplies ran low. With a brief meeting engagement against German strategic reserves, the tank corps assumed a defensive posture. It would attempt to hold its line of advance. With the tank corps, the commander creates a number of strongpoints, directs seizure of road junctions and coverage of likely routes of enemy withdrawal. He also establishes a corps mobile reserve and directs his brigade commanders to do likewise. This gives the units the potential to repel unexpected attacks and destroy infiltrating groups between the strongpoints.

While sitting astride the retreat routes of the German front units, the tank corps assumes a defensive line. Waiting for the relief by advancing front-line Red riflemen, the tank corps commander could delve into his years of battle and compare the phases of the war. By 1944, it seemed to him that they had mastered the art of war. The number of combat tasks that they had accomplished on this operation was considered only natural, but if they were transferred mentally back to the initial period of the war, they would be considered incredibly difficult, bordering on the impracticable. Such was the legacy in tank operations.

Conclusions

Reflections similar to the tank corps commander's have continued as current Soviet military researchers rigorously analyze their mobile group operations on the eastern front. Soviet analysts identified the same areas, i.e., command and control, logistics, air defense, fire support, intelligence, as have the Western speculators, on the problems of OMG-type operations. For the Soviets, in true Marxist-dialectical form, their solutions incorporated new technology with mobile group experience to allow for realistic consideration of the OMG role. The mobile group experience is an invaluable legacy to understanding similar operation on the modern battlefield.

NOTES

1. See C.N. Donnelly, "The Soviet Operational Manoeuver Group: A New Challenge for NATO," *International Defense Review,* 15(No.4, 1982), pp. 1177-86; C.J. Dick, "Soviet Operational Manoeuvre Groups: A Closer Look," *International Defense Review,* 16 (No. 6, 1983), pp. 669-776; John G. Hines and Phillip A. Peterson, "The Warsaw Strategic Offensive: The OMG in Context," *International Defense Review,* 16 (No. 10, 1983); Chris Bellamy, "Antecedents of the Modern Soviet Operational Manoeuvre Group (OMG)," *RUSI,* 129 (September 1984); Michael Ruehle, "The Soviet Operational Maneuver Group: Is the Threat Lost in a Terminological Quarrel?" *Armed Forces Journal,* 122 (August 1984); Henry S. Shields, "Why the OMG?" *Military Review,* 65 (November 1985), pp. 4-13j; Richard N. Armstrong, "Mobile Groups: Prologue to OMG," *Parameters,* Vol XVI No. 2, Summer 1986.

2. V. Odintsov and V. Ovsyannkov, "Rear Support for Mobile Groups," *Voyenno istoricheskii zhurnal* (Vizh), (No. 3, 1983), pp. 43-49; I. Tormozov and V. Tokarskii, "The Organization of Air Defense for Mobile Groups of Armies and Fronts During

Offensive Operations od the Second and Third Periods of the War," *Vizh,* (No. 4, 1983), pp. 21-27; A. Krupchenko, "Technical Support of Tank and Mechanized Corps Operations as Mobile Groups," *Vizh,* (No. 6, 1982), pp. 27-33.

3. N. Kireyev, "From the Experience of the Employment of Forward Detachments of Tank (Mechanized) Corps," *Vizh,* (No. 9, 1982), p. 21. The former Marshal of Soviet Armored Forces and Commandant for the Malinovskii Tank Academy, O.A. Losik, won his recognition and Hero of the Soviet Union Gold Star as a forward detachment commander. His tank brigade, in the capacity of a forward detachment, was the first to reach Minsk during the Belorussian operation of 1944. See note 11 below.

4. Ibid., p. 23; See also F.D. Sverdlov's *Forward Detachments in Combat,* Moscow: Voenizdat, 1986.

5. V.I. Chuikov, *The End of the Third Reich,* Moscow: Izdat "Sovetskaya Rossia," 1973, p. 80.

6. D.D. Lelyushenko, *Moscow-Stalingrad-Berlin-Prague,* Moscow, Nauka, 1973, p. 255.

7. P.P. Torstukha, R.M. Portugal'skii, *Troop Control in an Offensive,* Moscow: Voenizdat, 1981, p. 198.

8. Ibid., p. 178.

9. Ibid.

10. I.S. Konev, *Forty-five,* Moscow: Voenizdat, 1966, p. 106.

11. O.A. Losik, *Formation and Combat Use of Soviet Tank Troops During the Years of the Great Patriotic War,* Moscow: Voenizdat, 1979, p. 331.

CHAPTER 4

Mobile Groups: Prologue to OMG

The aim of Soviet military organizations is the destruction of the enemy. This destruction is to be accomplished by developing tactical into operational, and subsequently strategic, success. Like most art, this rather simple objective belies the technical difficulties of structure and execution. With significant changes in the past few years, Soviet armed forces have exhibited a surprising degree of flexibility and innovativeness in their force structure and fighting doctrine. In particular, the identification of the operational maneuver group (OMG) has been the focus of great speculation in the Western press.[1] While some articles have acknowledged Red Army mobile groups operations of World War II, most have failed to present a full appreciation of the legacy that mobile group operations represent to current Soviet Army military planners on OMGs. Concurrent with Western writings on the OMG, Soviet military publications, in a characteristically direct usage of their military historical studies and combat experience analysis, are defining the lessons of mobile group operations for applicability to OMGs.[2] Although World War II mobile groups operated in a comparatively simple environment - tanks sought to dominate the battlefield and aviation gave rudimentary support -a similar concept on today's battlefield would be dependent upon past lessons fused with modern technology and operational objectives. A closer scrutiny of mobile group operations on the eastern front may more sharply define our expectations in the structure, composition, and use of contemporary OMGs, for the OMG story cannot begin without the mobile group experience.

Before looking at the Red Army use of mobile groups on the Eastern Front, it is necessary to review the doctrinal context within which mobile groups developed. Soviet miliary research between the world wars concluded that in World War I breakthrough of the tactical defense zone in most cases did not lead to great operational success. Failure generally resulted from two basic causes. First, forces and equipment that could achieve high attack tempo were unavailable. Tanks and aircraft had shown their potential when used in small numbers. Their technical reliability, however, did not suit them to participate in high-speed, deep attacks. Second, the lack of improvements in breakthrough techniques and the unskillful coordination of separate arms, especially infantry and artillery, did not resolve the problem of breaking through an enemy defense to its entire depth.

Based on a rigorous analysis of World War I in his book, *The Character of Contemporary Army Operations*, V.K. Triandafillov, Chief of Operations, Red Army Staff, developed the concept of a mass mechanized army which could be employed deep into the battlefield. The Red Army's new, mechanized-based offensive doctrine began to take shape in the field service regulations of the late 1920s. Soviet military theoreticians reached the conclusion that successful offensive operations would require not only a decisive penetration of the enemy defenses, but also a rapid transfer of the main efforts of the attacker to the enemy's operational depth for the purpose of developing the offensive. Tank and mechanized forces would play the key role in performing the second mission.

Throughout the 1930s under the watchful eye of the youthful Marshal M.V. Tukhachevskii, the Red Army applied these theories through tactical maneuvers into force structural changes. As a result, by 1936, the Red Army had created four mechanized corps. The production of tanks and vehicles necessary to implement these theories became an integral part of the Soviets' first five-year economic plan.

On the eve of World War II, after the purge of the Soviet officer corps and confusing combat experiences from the Spanish Civil War, occupation of eastern Poland, and war with Finland, the Red Army leadership temporarily dismantled its mechanized corps. Only in late December 1940, based on General G.K. Zhukov's experience at Khalkhin Gol against the Japanese Army on the Mongolian-Manchurian border and observations of German Army success in France, did the Red Army leadership reverse their regressive trend and begin rebuilding a mechanized force capable of fulfilling a deep-battle concept. It was a reorganization caught mid-stride by the German Army invasion of June 1941.

During four years of fighting on the Eastern Front, the Red Army gradually applied their theory of deep combat operations. They solved the World War I problem of penetrating defenses by employing massed artillery fire and air strikes to support infantry making the initial penetration. Mobile groups followed on the most important axes with the objective of rapidly developing the attack to the whole depth of German defenses. Red Army experience convinced them that the decisive condition for complete destruction of the enemy was achieving a high attack tempo, for even short halts gave the enemy breathing space to maneuver or counterattack.[3] The Germans conditioned the Soviets well by their ability to react literally overnight to Red Army maneuvers.

The basic mission of mobile groups was to develop the tactical success. They were committed through gaps, at boundaries, or from the flank of the first-echelon units, primarily along successful axes. The axis for committing second-echelon units into battle varied according to situation. An important difference between mobile groups and second echelons was that mobile groups had specific missions and were committed to battle at or near the beginning of the operation in order to develop the attack swiftly in depth, while the second echelon was usually used after the immediate objective was secured.[4]

Experience showed that successful offensives required not only a decisive breach of the German defense but also a swift shift of the attacker's main effort into the enemy's operational depth for the purpose of developing the offensive. The key role in achieving the latter objective was played by the tank armies, cavalry-mechanized groups, and tank or mechanized corps, which were the mobile groups of armies and fronts.[5]

Under the scrutiny of the Soviet Supreme High Command, the Red Army General Staff closely studied tank armies and corps actions in all operations and generalized that experience. These experiences were distilled in the form of High Command orders and directives issued with important instructions for use of tank troops by armies and fronts.[6]

In the early months of the war after the destruction of the mechanized corps, fronts and armies did not possess large armored units. Front mobile groups consisted of cavalry and infantry supported by small tank battalions or brigades. This experience gained in the war with Finland when units, like mobile groups, were created with tank brigades reinforced by rifle battalions and sappers.[7] These groupings, however, had insufficient combat power to develop success to a great depth.

Coincidentally with the three periods of the war were three organizational phases which Soviet armored forces went through.[8] During the first phase, from July 1941 to early 1942, large tank units were pared down to smaller ones (brigades, regiments, battalions) for easier command and control of the limited number of available tanks. In the 1942 winter battles around Moscow, Red Army forces were unable to surround large enemy groupings and press home the deeper attack. Without large tank formations in the Front and army organization, Red Army leadership concluded that the important task of developing tactical success to the operational depth could not be fully achieved.

In the second phase of the war, from April 1942 to the end of the year, the Red Army formed tank and mechanized corps and, concurrently, their first tank armies. However, the tank and mechanized corps lacked an armored infantry, and the tank armies were formed from diverse infantry, cavalry, and tank units. Meanwhile, Red Army combat experience was also refining their offensive techniques. On main assault axes the Soviets began using large tank formations with the primary mission of assisting infantry forces in the breakthrough and development of the attack. The Red Army considered tanks primarily an offensive resource.

Initial offensive experiences with the newly formed 5th Tank Army at Voronezh in July 1942 revealed too great a disparity in combat capabilities and mobility among the army's rifle divisions, tanks and cavalry corps. Such a mobile group composition proved too awkward and unwieldy to control. The need to counter the considerable tank capabilities of the Germans spurred the Red Army to develop even larger, better balanced tank units.

In the third phase of development, from 1943 to the end of the war, the Soviets refined the organization and improved the composition of tank formations. In 1944, improvements in the firepower and maneuverability of tank corps increased their tactical independence. These improvements resulted in the creation of real capabilities for inflicting deep strikes against the Germans and for conducting combat operations with a higher momentum than before.[9]

By the end of the war, Red Army tank armies were composed of two or three tank or mechanized corps. The basic trend was toward increased firepower, greater unit autonomy, higher mobility and maneuverability, and more easily controlled regiments, brigades, and corps. These were characteristics necessary for successful mobile group operations.

The main operational objectives of mobile groups were to inflict deep strikes in order to scatter and destroy the defending units, to complete a tactical encirclement of the main enemy groups, to hold off enemy concentrations of tactical and sometimes strategic reserves, and to occupy and hold important objectives and lines until the approach of infantry units.

Successful operations by Front and army mobile groups in developing the attack to the operational depth depended greatly on circumstances in their commitment to battle, on close cooperation with other ground forces and air support, and on firm troop control. Ideally, the basic method for developing a breakthrough was to have mobile groups begin their combat operation only after the infantry divisions had broken through the German's main defensive zone. In most Red Army operations, however, mobile groups were committed on the first day of the battle within the German tactical defense zone. Mobile group tank strengths proved necessary to assist infantry units in creating a breach in the main defense zone. Only in isolated instances did the Red Army commit mobile groups to battle after the first echelon armies had already broken through the entire tactical defense zone. Usually, mobile groups were not intended for the breakthrough but rather for their specific mission.[10] The commitment of mobile groups to action on the first day was one of the decisive moments in the Front or army attack.

Operations of separate tank or mechanized corps acting as army mobile groups on the first day of the commitment to the breakthrough advanced from 15 to 50 kilometers. If the German defense was more deeply echeloned, the advance was held to 15 or 25 kilometers; if not deeply echeloned, the advance could be 35 kilometers or more.[11]

Mobile groups typically operated separately from the main forces of the front when they carried out their combat missions. Throughout the war the organizational structures of tank and mechanized formations that comprised the mobile groups were improved to ensure their combat independence and survivability when operating deep. The major lesson the Red Army learned was that separate mobile groups could be established and could function independently during operations when separated for long periods from the main attack forces.[12]

Mobile groups used concentration and assembly areas prior to commitment to combat. While corps-size mobile groups used concentration areas 25 to 150 kilometers from the line of contact, on an average concentration areas were 50 to 70 kilometers to the rear.[13] Concentration areas less than 50 kilometers from the German line were hard to conceal and hindered immediate redeployment to new or unanticipated axes. Areas more than 100 kilometers distant resulted in excessive expenditures in combat vehicles because of routine equipment malfunction on long-route marches before combat operations.

Army-sized mobile groups usually combined concentration and assembly areas which were usually outside population centers in natural cover and concealment. Assembly areas were designated for the transfer of mobile groups in the direction of impending operations, and they provided concealment as close as possible to the line of contact. These assembly areas were

20 to 40 kilometers from the line of contact for tank armies, 10 to 20 kilometers for corps.

Mobile group units were configured in assembly areas for ease of movement to a line of commitment without further regrouping. For its move to commitment, each corps was given two routes within a zone of 8 to 20 kilometers for tank armies, 6 to 8 kilometers for tank and mechanized corps.

Mobile group formations usually consisted of two echelons, artillery groups, and a reserve, depending on the situation and mission. In corps-size mobile groups at least two brigades (usually tank) and a large portion of the support weapons were assigned to the first echelon and designated to achieve the immediate mission. The corps second echelon reinforced the attack for deep exploitation, repelled German counterattacks, and consolidated success. Reserves, in battalion to brigade size, dealt with unexpected German tank attacks on the mobile group's flanks and rear, and exploited and consolidated offensive success.

The reconnaissance elements of first echelon brigades began the advance on their respective axes. Behind reconnaissance followed movement support detachments, which began necessary obstacle clearing on the movement routes for their parent brigade.

Upon approaching the line of commitment, forward detachments in the form of reinforced tank or mechanized brigades usually were deployed into combat formation and cooperated with the infantry units' attacks. Facilitators of battle, these forward detachments generally activated the offensive and immediately influenced the depth of the operation while in some measure disorganizing the defense. Having assisted in breaking down resistance in the main defensive zone, forward detachments rushed onto the second defensive zone in order to break through it on the march.

If mobile groups had to complete the breakthrough of the enemy tactical defensive zone after infantry units engaged the main German defensive zone, then in this case only, forward detachments or part of the mobile groups's first echelon forces were called upon to seize the second German defensive line. Only in those instances did combat actions in breaching the second defensive line of the Germans take on a drawn-out nature.

Forward detachments, usually a reinforced tank brigade, had the specific mission to seize and hold important tactical or operational objectives, disrupt German defenses in depth, and partially prevent maneuvering of enemy operational reserves.[14]

Under favorable conditions, mobile groups were committed to the battle after the first echelon armies had already penetrated the entire German tactical defense zone and sometimes even into the rear area. Most often this occurred on the second through fifth day of the operation at a depth of 15 to 35 kilometers from the forward edge of the battle area and sometimes even farther. Such a method of commitment to battle was called a "clean" breakthrough.

If the mobile groups were committed to a clean breakthrough, they usually advanced from the assembly areas to the line of commitment with the forward detachments of the first echelon units in front of them. The mobile group main force moved behind the lead units in route columns and, after passing through the entire tactical defense zone of the Germans, overtook the infantry and rushed into the operational depth. Sometimes only the forward detachment of the mobile groups deployed into battle formations when overtaking the infantry.

Aviation formations selected to support the combat actions of the mobile groups became operationally subordinate to the commander of the mobile group with the beginning of commitment to battle. To coordinate the actions of ground attack aircraft, the commander of the supporting air division was located in the mobile group command post and had communications with both the airfield and aircraft that were airborne. In the first echelon brigades and forward detachments were air army personnel who would guide aircraft to specific targets. Mobile groups seized serviceable airfields on which supporting aircraft could be quickly relocated in order to keep air support in the immediate vicinity of mobile groups.[15] Air strikes combined with artillery fires cleared the path along which the mobile groups operated.

After completing the breakthrough of the German defense, the most characteristic missions of mobile groups operating in the operational depth were to engage German reserves, preempt subsequent defensive lines, cross water obstacles, fight to hold key terrain, pursue withdrawing German forces, and seize important installations or facilities.[16]

Once through the tactical defensive zone, the mobile group's main mission was the defeat of the German's operational reserve.[17] Advancing at high rates in the operational depth well ahead of friendly first echelon forces attracted German reserves. Mobile groups would immediately encounter reserves which rushed to close the breakthrough and intended to prevent the advance by holding them within the limits of the tactical defense zone. Impressed by the Soviet speed in exploiting even the smallest penetration, German commanders were forced to make every possible effort to immediately seal off penetrations, however small their own counterattacking force.[18]

Combat actions against attacking reserves took the form of meeting engagements. These engagements would develop in broad zones of 20 to 60 kilometers and last from one to three days.[19]

The Soviet military researchers concluded that in the course of fighting in the tactical defense zone German commanders, as a rule, used all of their closest operational reserves and for the next several days were practically unable to offer any serious resistance to the offensive of mobile groups.[20] German units then rushed to bring reserves from a strategic depth into the offensive zones of mobile groups and by counterattacks or by occupying defensive positions delayed further advances by Red Army mobile forces.

Soviet analysts noted that conditions for fighting German reserves in the beginning and at the end of an operation differed substantially. Fighting German reserves during commitment to battle, although it did affect the future course of the operation, nevertheless usually was fought jointly with the first echelon units and in cooperation with aviation. Mobile group combat vehicles and equipment were close to established strengths, and they usually achieved a force ratio in Red Army favor.

Conditions for fighting German reserves in the closing phase of an operation differed completely. By then, mobile groups had expended a considerable part of their forces and weapons and were deficient in supplies and fighting vehicles. Moreover, they operated a great distance from the main forces of the fronts and sometimes without air support. While in the first phase of the operation Red Army mobile groups fought German unit reserves by direct offensive actions, in the closing phases fresh German reserves frequently forced mobile groups onto the defensive after brief meeting engagements.

The mobile groups' ability to develop the tactical success was directly dependent on their adeptness to overcome rear defensive lines and water obstacles quickly. After completing a breakthrough of the Germans' tactical defense zone and defeat of their immediate operational reserves, mobile groups turned to pursuit of retreating German troops. An important requirement during pursuit was a high offensive momentum. Tempo was the nexus between tactical capabilities and operational possibilities. Soviet studies showed that the tempo in the operational depth largely relied upon the skillful and dynamic actions of forward detachments.

Forward detachments rushed into the depth of enemy defenses, slipped into the rear of their retreating groups, captured bridges and river crossings, cut off the retreat of the enemy troops, and also forestalled their occupying defenses at earlier prepared lines. In the event the mobile group encountered major German strongpoints, they tried not to engage them but bypass these points to maintain momentum and move even deeper.

German defenses in the operational depth usually were characterized by an insufficiently developed fire plan, hastily organized and spread units, and poorly used terrain marked by incomplete engineer preparations of barriers and obstacles. Against these hastily occupied defenses, the mobile group broke through on the move or with a short artillery preparation when an attack on the move was unsuccessful. In the latter case, the breakthrough's failure was often attributable to mobile group commanders waiting for the arrival of additional support, usually artillery. By hesitating they lost the element of surprise, and German units had time to consolidate and improve their defense. "Achieving dependable artillery support of the operations of tank armies in operational depth," a senior Soviet doctrine writer noted, "required the creation of a sufficient quantity of fast moving artillery . . . [To] the very end of the war this was not done."[21] A number of mobile group operations successfully breached defenses in the operational depth on the move without massed artillery by substituting the firepower of the tank cannons and the support of air strikes.

Organizing and preparing breakthroughs on the move was done during the approach to the defensive line. The mobile group's offensive zone width varied and depended on the availability of men and equipment, the German grouping, the nature of the defense, and terrain conditions. Correlating these factors allowed Soviet commanders to concentrate their main efforts. During attacks, mobile groups often were not superior overall to German forces. However, concentrating men and equipment on the main axis, the Red Army achieved a considerable local superiority.

The simultaneous breakthrough of German defenses in several sectors proved to be quite effective. Large Red Army offensives often used a multiple of mobile groups advancing along parallel axes. Constant cooperation between the neighboring mobile groups was accomplished through mutual awareness of missions and the constant exchange of information. Such a method of attack did not allow the German commanders to maneuver their forces and enabled Red Army forces to split, isolate, and destroy defenders in a short time. Such a breakthrough technique increased the difficulty for defenders to detect the main strike axis and gave them little opportunity for organized resistance. Additionally, spreading mobile groups avoided a large concentration of men and equipment, which lowered losses from German ground attack aircraft.

In engagements for defended built-up areas, mobile group momentum decreased sharply and many times stopped entirely. Therefore, the mobile groups tried to bypass built-up areas and attempted to cut off and encircle German garrisons for subsequent destruction by following infantry units. While it was advisable to go around German centers of resistance, the Soviets did take into consideration whether more time would be spent bypassing one than it would take to deploy and organize an engagement to overcome it.

Whether centers of resistance were bypassed by mobile groups in the operational depth also depended on the nature of their mission. In the course of some operations Germans blocked the path of mobile groups with an organized defense on a wide front, occupying it with individual strongpoints. In these situations, mobile groups were forced to deal with the German resistance, breach the defense, and then resume their freedom of operational maneuver.

Quite often mobile groups were forced to cross several water obstacles during an operation. In fact, tank armies had insufficient river crossing equipment. What was assigned to them often lagged behind because it lacked comparative cross-country mobility. Thus one of the basic missions of forward detachments was to seize river crossings, which made possible a more rapid crossing by the mobile groups.[22] Success depended on reconnaissance in order to determine the number of bridges and fords. Forward detachments were given the mission of seizing identified crossing points. While artillery moved forward as fast as possible, aviation support had the mission of isolating the crossing sector from German reserves and establishing air supremacy in the crossing area.

In the course of offensive operations, successful night operations were conducted by forward detachments of mobile groups. Regrouping and marching by the mobile groups were done mainly at night. This enabled them for a certain amount of time to conceal forces from the Germans, the plan of the Soviet command, and sharply reduced losses from German air strikes.

The Red Army Belorussian offensive operation of 23 June to 29 August 1944, depicted on the accompanying map, is a good example of extensive and multiple uses of mobile groups by armies and fronts. In an effort to complement the Allied invasion at Normandy, the Red Army planned a major offensive operation with the objective to destroy the German Army Group Center and parts of Army Groups North and Northern Ukraine.

The tank armies and mechanized corps (the latter were included in the cavalry-mechanized groups [CMG]) were used primarily as Front mobile groups. In the Third Belorussian Front, two mobile groups were organized: 5th Guards Tank Army and a CMG (III Guards Cavalry and I Mechanized

Corps). The Second Belorussian Front's mobile group consisted of two tank brigades, a rifle division, and a self-propelled artillery regiment. An exception, the I Tank Corps, was assigned as the Front mobile group for the First Baltic Front.

The separate tank corps (with the exception of the I Tank Corps, First Baltic Front) were employed as mobile groups for combined-arms armies advancing in the Fronts' main sectors. The II Guards Tank Corps comprised the mobile group of the 11th Guards Army; the IX Tank Corps did the same in the 3rd Army; I Guards Tank Corps, in the 65th Army; and XI Tank Corps, for the 8th Guards Army.

A unique employment of the mobile groups in the Belorussian operation was that separate tank corps, as army mobile groups were used for completing the breakthrough of the tactical defense zone and for supporting the commitment of the Front mobile groups, tank armies, and CMGs. The So-

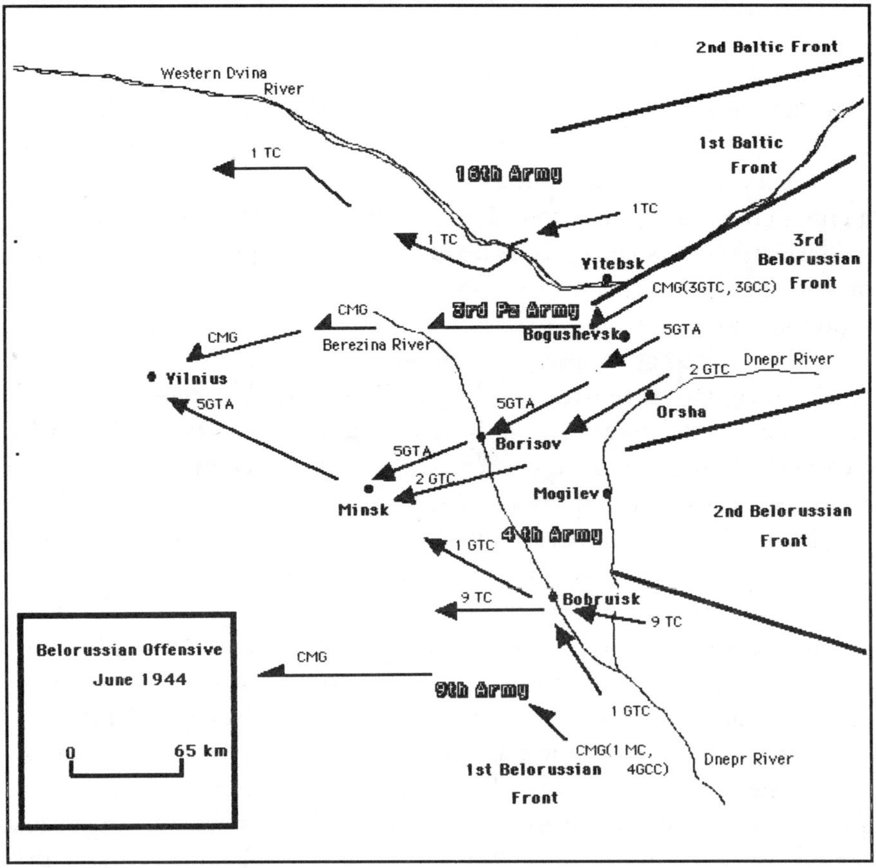

viets planned the successive commitment of the mobile groups to battle. For example, the II Guards Tank Corps, 11th Guards Army mobile group, was to be committed to battle on the first day of the operation, while on the fourth day in the same sector the 5th Guards Tank Army, the mobile group of the Front, was to be committed.

The commitment to battle of the mobile groups of the armies and Fronts was a crucial and difficult stage. This was marked by a number of essential features determined by the overall plan of the operation, by the nature of the German defense, and by terrain conditions. Depending upon the developing situation, the commitment to battle of the follow-up echelon was carried out on different axes and frequently in a manner quite unlike what was envisaged in the plans of the army and Front operations. For example, the 5th Guards Tank Army was to be committed to battle according to the first variation, that is, on the Orsha-Borisov axis; but this was canceled as the German defenders put up stubborn resistance to troops of the 11th Guards and 31st Armies. The 5th Guards Tank Army was regrouped to the left wing of the 3rd Belorussian Front and committed to battle on the Bogushev-Borisov axis, where a CMG had been committed two days previously. Having caught up with the rifle formations at a depth of 30 kilometers from the previous forward edge, the tank army began to rapidly pursue withdrawing German units and came out in the direction previously planned for it.[23]

Red Army forces broke through German defenses simultaneously in six sectors, encircled and destroyed German forces in the areas of Vitebsk and Bobruisk, destroyed German units in Orsha and Mogilev, and then encircled and destroyed the main forces of Army Group Center in the area of Minsk. Red Army forces achieved the liberation of Belorussian, seized parts of Lithuania, Latvia, and Poland, and forced the Narev and Vistula rivers, having advanced 550 to 600 kilometers.

Soviet analytical conclusions on experiences of mobile group combat operations within the operational depth ascribe their achievements to swift, decisive, and continuous day and night action and to the use of wide maneuvers when carrying out combat missions. The high momentum of mobile group attacks was achieved mainly through their aggressive use of forward detachments. They accomplished decisive action by avoiding drawn-out engagements and battles for major strongpoints unless necessary. By maintaining continuous operations they attempted to enter the Germans' rear area and to intercept their main lines of communications as quickly as possible.

As the Red Army refined their employment of mobile groups during the second and third periods of the war on the Eastern Front, many aspects of combat and service support for these groups were continuously improved in capabilities and numbers.

A significant area relevant to OMG employment today was the organization and employment of air defense for mobile groups. Through the course of the war the primary trend was a greater density of ground antiaircraft fire under a centralized control and closer cooperation with fighter aviation support.[24] Antiaircraft artillery groups were formed and organized to protect mobile groups from assembly areas through the breakthrough and into the operational depth. Fighter aviation from the front air armies flew counter air protection to cover the mobile groups throughout the operation against German airstrikes.

Red Army studies of air defense support to mobile groups identified the following apparent needs: a highly mobile, armored antiaircraft weapon system that had cross-country and adverse weather capabilities; an organized system of reconnaissance for enemy air, using mobile radars in the course of the operations; and a centralized control of air defense personnel and weapons in repelling enemy air raids.[25]

Rear services for mobile groups proved to be the most difficult of organizations and operations to support.[26] Mobile groups consumed large quantities of fuel and ammunition: "According to World War II experience the average daily fuel and ammunition requirement of a tank army of medium strength was 600-700 tons. To satisfy this requirement it was necessary to provide, for a distance of 200-300 kilometers, 270-300 trucks with loads."[27] Front transports delivered comparatively little fuel to mobile groups. Mobile groups were forced to organize their own supplies for the entire offensive operation, frequently reducing their rate of advance.[28]

Attempts were undertaken by the Fronts to supply mobile groups by air transports, but a small number of available air transport carried an insignificant amount. One notable success was the major air resupply of forward detachments of the 6th Guards Tank Army during the Manchurian campaign in August 1945.

Local food resources were widely foraged to preserve the precious cargo space for more critical ammunition and fuel. The nature and amount of work related to road traffic support necessitated having such road units as a bridge-building company, traffic control platoon, road platoon, and security platoon.

Mobile groups used all forms of transport to evacuate sick and wounded. When roads were lacking or Germans cut lines of communications to the rear, sick and wounded were left at temporarily organized hospitals near inhabited areas until the arrival of front line forces.

Because of the high rate of advance and great distances from main forces of Fronts, the importance for autonomy in rear services for mobile groups increased. A recent Soviet Army review of mobile group experiences concluded, "A profound and creative study of the question related to rear services support for mobile groups in the course of the last war will aid in successfully solving today's problems in developing organic and operational rear services."[29] Such is the Soviet advice for what may be assessed as the Achilles' heel for OMGs.

The mobile groups' operational tempo and distance from friendly front lines significantly influenced maintenance support. Combined arms armies did not have organic repair facilities until 1944, when they began to include two mobile repair bases for medium repair on tanks and motor vehicles.

Maintenance support for mobile groups was organized for three stages in an operation: preparation, movement to the line of commitment, and operations in the operational depth.[30] The Red Army brought units to full, or near full, strength in the preparation phase and attempted to maintain that level in movement to commitment. The third stage, mobile maintenance support in the operational depth, was, of course, more problematic. Studies showed the average daily breakdown rate of tanks and self-propelled assault guns in corps-sized mobile groups to be 8 to 10 percent from start to combat, and in two to three days of battle they lost up to 50 percent of their tanks.[31]

The basic method was to repair equipment directly in the forward unit areas. Repair and recovery groups of tank battalions deployed behind battle formations of battalions while maintenance support companies of brigades and corps deployed behind first echelon brigades. Maintenance observation posts were organized for checking tanks and summoning repair equipment to damaged vehicles in tank battalions. Brigade maintenance teams repaired directly at the place where equipment broke down. Vehicles requiring labor intensive repairs were evacuated to the corps assembly point for damaged vehicles. When rates of advance were high (40 to 50 kilometers) or limited recovery capabilities existed, corps did not organize assembly points. Maintenance support for corps mobile groups was based on organic resources located at the rear command post.

For corps-sized mobile groups, army and front maintenance resources were used to repair combat vehicles on the line of commitment to a breakthrough to ensure the complete moving of tank and mechanized corps' organic repair and recovery units into the operational depth.

The quality of maintenance support also was affected by the supply of spare parts and materials. Centralized supply from army and Front depots depended on ground transport lines. When these lines were broken, maintenance units had to use parts and assemblies from vehicles that would not be repaired. Mobile group experience showed that corps-sized mobile groups operating in the operational depth usually received 40 to 50 percent and sometimes all of their parts and assemblies from irretrievably lost equipment.[32] German General Hasso von Manteuffel thought Red Army salvage and repair services performed extraordinary feats. He therefore issued orders that incapacitated Soviet tanks were to be set on fire.[33]

Another area of World War II mobile groups experience that bears significantly on OMG operations today is command and control. A command post (CP) system was established that included in the first echelon of an army (or corps) mobile group a main command post and a rear command post. During operations, operational groups headed by the army or corps commander often were established to accomplish the role of a forward CP.[34] Small in size, operational groups were assigned staff officers with communications means to direct army and corps staffs in accordance with the commander's instructions.

During the course of an army-sized mobile group operation, the forward CP was located six to ten kilometers from the line of contact; the main, 15 to 25 kilometers. The rear CP was located 25 to 30 kilometers behind the corresponding rear Cps of subordinate corps.[35]

Army-sized mobile groups changed CP locations by leap-frogging, rear CP to the main, main CP to the forward. These shifts were done at night during the last active periods of combat. By the middle of 1944, the established practice was to work out a special plan for the movement of CPs. Shifting CPs depended on the tempo and conduct of the attack. Forward CPs moved once, twice, or even more times a day; main CPs usually moved once a day; and rear CPs moved once in two or three days.[36]

A corps-sized operation was very mobile. For example, Soviet analysts computed that the IX Tank Corps during the Vistula-Oder operation halted twenty-thirty minutes for every one to five hours of movement. The speed of set-ups and tear-downs of CPs depended largely on staff teamwork and quick preparation of communications and equipment: corps CPs averaged forty-sixty minutes.[37]

When distances were too great for available radios and made dependable direction and control impossible, the Red Army turned to mobile means, such as communications planes. In these circumstances, forward CPs were frequently established.[38]

During rapid rates of advance, commanders issued short combat orders by radio and liaison officers carried duplicate messages. Sometimes at the end of the day, subordinate commanders and their chiefs of arms and services would meet in the main CP. They would summarize the day's combat operations, adopt decisions for the following day, and assign the combat missions.

The primary purpose for mobile groups was to rapidly shift the focus of combat to the rear area and thereby maintain a high rate of advance; today's OMGs have the same purpose. The importance of this dimension of operational-level warfare was succinctly described by the well-known Red Army Front commander Marshal I.S. Konev, who wrote in his memoirs: "Thus the whole orderly system of the enemy's defenses which envisaged an appropriate sequence of committing the reserves to action, was disorganized. And this was very important, because precisely such a disruption of the integrity of the enemy's forces and of the system of their control is a sine qua non for successfully developing an operation to great depth."

With a vast reservoir of available historical data, the Soviets believe they have the basis for operational analysis that can adapt the previous concept to the modern battlefield. With rigorous study in areas of operational doctrine, logistical and fire support and command and control, they will have considered critical mobile group problems relevant to successful OMG operations. Their adaptations also will incorporate an understanding of the effects of technology and its influence to the full depth of the enemy defenses with strikes by missiles, aviation, long-range artillery, and airborne and airmobile assaults. How the Soviets define and refine combat entities within the context of their historical experience will shape the force structure, equipment composition and employment of OMGs. Now, the OMG story begins.

NOTES

1. See C.N. Donnelly, "The Soviet Operational Manoevre Group: A New Challenge for NATO," *International Defense Review,* 15 (No. 9, 1982), 1177-86; C.J. Dick, "Soviet Operational Manoevre Groups: A Closer Look," *International Defense Review,* 16 (No. 6, 1983), 669-776; John G. Hines and Phillip A. Peterson, "The Warsaw Strategic Offensive: The OMG in Context," *International Defense Review,* 16 (No. 10, 1983); Chris Bellamy, "Antecedents of the Modern Soviet Operational Manoevre Group (OMG)," *RUSI,* 129 (September 1984); Michael Ruehle, "The Soviet Operational Maneuver Group: Is the Threat Lost in a Terminological Quarrel?" *Armed Forces Journal,* 122, (August 1984); Henry S. Shields, "Why the

OMG?" *Military Review*, 65 (November 1985), 4-13.

2. G. Chernykh, "On the Employment of the 6th Guards Tank Army in the First Echelon of a Front," *Voyenno Istoricheskii Zhurnal* (*Vizh*) (No. 3, 1981), pp. 66-69; N. Kireyev, "From the Experience of the Employment of Forward Detachments of Tank (Mechanized) Corps," *Vizh* (No. 9, 1982), pp. 20-27; I. Krupchenko, "Methods of Developing Success at Operational Depth by the Forces of Tank Armies, Tank and Mechanized Corps," *Vizh,* (No. 7, 1981), pp. 12-20; V. Odintsov and V. Ovsyannkov, "Rear Support for Mobile Groups," *Vizh* (No. 3, 1983), pp. 43-49; I. Tormozov and V. Tokarskii, "The Organization of Air Defense for Mobile Groups of Armies and Fronts During Offensive Operations of the Second and Third Periods of the War," *Vizh* (No. 4, 1983), pp. 21-27; A. Krupchenko, "Technical Support of Tank and Mechanized Corps Operations as Mobile Groups," *Vizh* (No. 6, 1982), pp. 27-33; see also for major studies of armored forces: *O.A. Losik, Formation and Combat Use of Soviet Tank Troops During the Years of the Great Patriotic War* (Moscow: Voenizdat, 1979) and A.I. Radzievskii, *Tank Strike* (Moscow: Voenizdat, 1977).

3. A.I. Radzievskii, *Tactics by Combat Example* (Division), (Moscow: Voenizdat, 1976), p. 71.

4. I. Krupchenko, p. 12.

5. Ibid.

6. Losik, p. 118.

7. P.A. Rotmistrov, *Steel Guard* (Moscow: Voenizdat, 1984), p. 45.

8. Soviet recent historiography, which calls the fighting between the Red and German Armies the Great Patriotic War, has divided the war on the eastern front into three periods. The first period, 22 June 1941 to 18 November 1942. The second period, 19 November 1943 to the end of 1943, is the basic turning point in the course of the war. The third period, January 1944 to 9 May 1945, the defeat of the German forces and the liberation of occupied European countries.

9. Losik, p. 72.

10. Ibid., p. 120.

11. Ibid., p. 122.

12. P. Kurochkin, "Operations of Tank Armies in Operational Depth," *Voyennaya mysl'* (November 1964), translated in *Selected Readings From Military Thought 1963-1973,* compiled by Joseph D. Douglas and Amoretta Hoeber, published under the auspices of the U.S. Air Force (Washington: GPO, 1982), p. 65.

13. Losik, p. 123.

14. Modern battle missions for forward detachments could include the above missions with the addition of destruction of the enemy's nuclear delivery systems, small advancing enemy reserves, and air defense groups. In the April 1965 issue of

Voyennaya mysl', General Vorob'yev noted that it would be reasonable to adopt the term "Operational Forward Detachments." Both the old and new concept is to dispatch detachments with the "desire to increase the speed of the advance and to increase the depth of simultaneous action against defenders on important axes." I. Vorob'yev, "Forward Detachments in Offensive Operations and Battles," *Voyennaya mysl'* (April 1965), translated in *Selected Readings From Military Thought* 1963-1973, pp. 97-98, 101.

15. Losik, p. 141.
16. I. Krupchenko, p. 13.
17. Kurochkin, p. 75.
18. Department of the Army Pamphlet 20-233, *German Defense Tactics Against Russian Breakthroughs,* October 1951, p. 32.
19. I. Krupchenko, p. 16.
20. Losik, p. 137.
21. Kurochkin, p. 79.
22. Ibid., p. 74.
23. O.A. Losik, "Application of Armored and Mechanized Forces in the Belorussian Operation," *Vizh* (No. 6, 1984), pp. 22-23.
24. I. Tormozov, p. 27.
25. Ibid.
26. V. Odintsov, p. 43.
27. Kurochkin, p. 80.
28. V. Odintsov, p. 43.
29. Ibid., p. 49.
30. A. Krupchenko, p. 28.
31. Ibid., p. 31.
32. Ibid., p. 33.
33. B. H. Liddell Hart, *The Other Side of the Hill* (London: Cassell, 1951), p. 333.
34. I. Krupchenko, p. 18.
35. Radzievskii, *Tank Strike,* p. 205.
36. Ibid., p. 208.
37. P.P. Torstukh, R.M. Portugal'skii, *Troop Control in an Offensive* (Moscow, Voenizdat, 1981), p. 178.
38. Kurochkin, p. 80.

PART II: CAPABILITIES

CHAPTER 5

Hunting Tongues

Fighting on the Eastern Front during the Second World War assumed an aggressive, bloody fanaticism pervading all aspects of combat to include tactical intelligence operations. Red Army commanders and staffs relentlessly searched for all clues to the disposition, activities, and intentions of opposing German forces. While use of aerial observation, reconnaissance-in-force, and radio intercepts all contributed to the gathering of valuable information, the prisoner of war often served as an important information source to complete the intelligence estimate on the current enemy situation. As the then 39th Army chief of intelligence, Colonel Maksim Afanas'evich Voloshin, wrote in his memoirs, ". . . a captured German soldier with his soldier's book had great significance."[1] However, the aggressive fighting on the Eastern Front did not allow the intelligence organizations, particularly in the Red Army, to wait for prisoners to fall into their hands as a result of a tactical operation, but rather they sought to acquire prisoners well before an offensive operation. This active acquisition of prisoners resulted in an interesting and often overlooked aspect of Red Army tactical intelligence collection.

Marauding Red Army reconnaissance patrols, called, "Hunts," (poiski) infiltrated German forward fighting positions and rear areas seeking to capture prisoners, documents, and, at times, inflicting losses on the enemy.[2] The "Hunt" patrol became a method by which tactical commanders conducted a surprise raid with the specific objective to capture a "tongue" (yazyk) - an information prisoner.[3] In these unique "hunt" patrols, the reconnaissance scouts were known as "hunters," and their search for "tongues" was playfully referred to as "linguistics" (yazykovedenie) - the science and study of tongues.[4]

The Red Army's increasing emphasis and goal for effective tactical reconnaissance became evident early in the war, and exemplified another dimension of the Soviet's warfighting capability that matured into a fine art through the war years. This maturity was the result of a systematic absorption of practical combat experience and its dissemination throughout the army. The Red Army General Staff established a method for collecting and processing its war experiences for lessons learned, and used published studies as a primary means for passing on conclusions. The second volume of *Collection of War Experiences for Study* contained a detailed section on the conduct of various tactical reconnaissance operations.[5] In seeking immediate, practical remedies, the study candidly identified shortcomings and gracelessly assigned responsibility for problems. "The absence of control over the work of reconnaissance elements naturally gives birth to deception and fraud, which was noted more than once in the formations of the 48th Army on the Bryansk axis. Having arrived at an enemy wire obstacle, the scouts often laid in waiting, and at dawn they returned with the standard report: "The reconnaissance was conducted without any results."[6]

The study explicitly blamed commanders and their staffs. "The fact that military reconnaissance up to now is at a low level is the fault not only of the commanders of reconnaissance units, who often do not have sufficient preparation and experience, but also the division and unit commanders as well as chiefs of staff and intelligence sections of higher headquarters."[7] While the requirements and tasks for "tongues" would come from the Army level, the execution was generally a division responsibility. But, in the early stages of the war, these tasks were passed along to the reconnaissance units without proper supervision and guidance from higher commanders and staffs.

The dangers and risks in this type of patrol cannot be overstated. No general analysis from Red Army experience has appeared on the success rate, but one suspects many attempts failed. Some of the difficulties in these patrols are evident in post-war writings. For example, the following German account appeared in a United States Army historical study on small unit actions during the German campaign in Russia.[8] The details of the account suggest that some of the Soviet leadership was captured.

The situation occurred during the last ten days of August 1941, a time when Soviet inexperience was clearly apparent. The German offensive on the southern front had lost momentum and stopped along the Dnepr river. Red Army units dug in along the east bank were still conducting limited counterattacks near Cherkassy. While an unidentified German infantry division prepared for a forced river crossing at Cherkassy, another division deployed along a wide front with the primary mission to protect the assaulting division's northern flank. In particular, the German 196th Infantry Regiment had the mission to defend the river line between Dakhnovka and Svidovok against Russian crossing attempts.

On the night of 21 August, local inhabitants swam the river to deliver information about the German dispositions to the Red Army rifle regiment on the opposite bank. Receiving this information, the Red Army regimental commander decided to send reconnaissance patrols across the river in an effort to gain more details. As an initial step, the commander reinforced his covering forces on the three islands in the middle of the Dnepr river. The covering forces would help protect the reconnaissance effort with careful observation by artillery observers. Upon completion of the mission the reconnaissance patrols would return to these islands. From his reserve battalion, the regimental commander selected forty men who appeared to be best suited for the task of gathering information behind enemy lines. Two reconnaissance detachments, each composed of one officer, three noncommissioned officers and sixteen men, were organized. Lightly equipped and armed each man carried only a submachinegun and three hand grenades, and wore a lightweight fatigue uniform and straw shoes.

On 23 August regimental orders were issued to the reconnaissance patrol. The regiment required specific information regarding location of positions and composition of German forces. The orders stated, "it is essential that some prisoners be taken and brought back for interrogation."[9]

At 0300 hours on 24 August the detachments proceeded to their jump off positions on the islands and remained there all day and into the evening. To avert premature detection the men carefully concealed themselves and their equipment, particularly their flat-bottomed boats. The departure time was set for 2200 hours which was after sunset on the long summer day. The entire operation was commanded by Captain Orlov. Lieutenants Novikov and Mirskii led separate patrols.

After dark, Lieutenant Novikov's patrol boats silently pushed off into the river from the northern end of the river island crossing the remainder of the river to the German shore. After several minutes all four boats landed on the western bank, which rose some 15 to 20 feet above the river line. Lieutenant Novikov climbed up, looked around, and found the immediate area unoccupied. He placed a wooden marker to indicate the concealed boat point. Lieutenant Novikov left a security detail of five men near the landing site. The rest of the patrol cautiously moved inland about a hundred yards before turning south.

Around 2300 hours the sound of approaching voices and footsteps was heard. The patrol quickly took cover. Four Germans passed close by. At a hand signal from Novikov, the Russians jumped the German soldiers and sought to disarm and capture them as noiselessly as possible. One of the Germans, however, succeeded in firing several bursts with his submachinegun while another shouted, "Russians, Russians!" In the ensuing hand-to-hand fighting, all four Germans were killed and one patrol member lay dead and four others wounded.

The Germans were alerted, and sounds of their approach could be heard from several directions. The Russian patrol quickly made its way back to the river. When they were only 50 yards from their goal, the river bank, German flares launched into the sky illuminating the area all around them and machineguns opened up at close range, forcing them to take cover. The patrol made a bold rush for the river. One Russian was killed by a machinegun burst. Novikov and the rest of his patrol reached the river, shoved off with their boats, and returned to the river islands. Despite the reconnaissance patrol's failure to snatch a "tongue," the Soviets did determine that definite gaps existed between the German positions. Roving combat patrols maintained a close surveillance.

Meanwhile, Lieutenant Mirskii's patrol had reached the west bank. He reconnoitered the immediate area of the river bank and found no Germans. Leaving the bulk of his force behind, Mirskii took five men and searched southward for German positions. After advancing about 150 yards he smelled the aroma of tobacco smoke. Mirskii cautiously crawled forward while the others prepared to cover him with protective fires. Within a few minutes he reached a low wire entanglement that appeared to be several yards in depth. Beyond the wire barrier, the Lieutenant heard very faintly the sound of conversation and the tapping of metal on metal.

Carefully Mirskii crept back to the five-man detail and then joined the rest of the party. After assigning five men to provide security from the south, he led the bulk of the patrol northward. After a short distance, a German machinegun suddenly opened fire from the vicinity of the landing point, which was now to the right rear of the patrol. Simultaneously, loud noises were heard as German flares illuminated the entire area. To avoid being cut off Mirskii quickly doubled back toward the landing site. The Russians soldiers quickly dispersed and tried to make their way back to the river. Intermittent firing from the south indicated that the security detail was still engaged by the German roving patrol.

Eleven members of the Russian patrol succeeded in reaching the river and assembled at a point to the north of the landing site where they waited patiently for the rest. The Germans carefully searched the area. An hour passed. The remainder of the patrol, including Lieutenant Mirskii and Sergeant Petrov, had still not returned, so Sergeant Rudin ordered two men to pull the boats upstream. They accomplished this under the protective cover of the high, overhanging river bank. The remnants of the patrol embarked and returned to the river island. This patrol, like Lt Novikov's, found wide gaps in the German lines closely watched by numerous patrols. Some of the German positions were not found in the locations reported by the local inhabitants. Like many of the early Red Army efforts in tactical reconnaissance, the small unit commands had thrown together patrols quickly with untrained soldiers. The unsurprising results were inconclusive reporting and unnecessary casualties.

Red Army General Staff study on war experiences criticized pro-forma actions by front and army staffs who did nothing more than process plans and papers. Placing the burden and responsibility on small unit commanders as the regimental example above illustrated was unacceptable. A simple reprinting of the higher headquarters orders and plans with no thought nor application to the actual situation by the subordinate commands, all contributed to poor results. The training of reconnaissance units was unorganized at best and commonly nonexistent. The ultimate remedy according to the study was, "The more rapid the commanders of formations and staffs at all levels transferred from paper methods of reconnaissance to direct work on the terrain with those involved, the sooner positive results will be seen."[10] This example of the need for specific involvement is a universal problem in all armies as their leadership transitions from peacetime theory to wartime practice. The realities of combat actions demand a commander's mastery of detail and persistence involvement in coordinating actions at all unit levels.

The success of reconnaissance patrols, particularly the hunt, depended directly on the time given to preparation, and "on how the commander who is organizing the reconnaissance thinks out and foresees questions of support of the hunt; it also depends on how successfully personnel for conducting the raid are chosen."[11] The war experience studies provided measures and actions to improve the chance for success in the hunt for tongues. Upon receipt of the mission, the commander, recommended the study, should accomplish the following steps for organizing a hunt.

The first step was a target assessment. The commander must choose the target of the hunt if it has not been indicated in the order from higher headquarters. Targets of the hunt should be individual soldiers, small groups, patrols, listening posts, scouts, or sentries.[12] Junior officers who might know more than the average soldier were considered prize catches. Knowledge of the routine of the chosen target, such as place and time of sentry shifts, meal time, and rest time played a large role.[13]

The target area had to be studied in detail. Approaches to target and exit routes had to be identified. Careful study of the deployment of the enemy, the location of his command and observation positions, headquarters, communication centers, and paths of communication precede the dispatch of "hunter" groups. Locating enemy fire points which would hinder the execution of the mission proved very important. Observation of the target took place for a long period of time (twenty-four hours or more) and included, above all, carefully organized observation. A study of the enemy's patterns of activity shaped the planned operation, such as the time and path of the patrol. Sometimes, the study recommended the necessity to send out small reconnaissance groups or individual scouts, including a sapper to reconnoiter the nature of the mine fields, signal wires, depth and width of water obstacles, and other types of barriers. Close, detailed observation identified the routes, times, and habits of the enemy sentries to allow them to be captured and at a time which would avoid discovery by supervisors or guard changes.

Hunts required a complete plan of action in which the commander should distribute his forces and means for attack and support; determine missions for supporting artillery, mortars, and medium machine guns; establish the time for commencement of activities and routes of entry and withdrawal upon completion of the mission; establish hand and arm signals for communications within the group and for calling supporting fire.

The commander was required to assign subordinates missions on the basis of the terrain and the compiled plan of action. He gave them adequate time for assigning the mission to everyone performing it and acquainting themselves with the terrain and target. Typical of the centralized Soviet leadership style, the commander personally verified the patrol members' understanding of the mission and checked the readiness of all forces and means to execute and support the operation.[14] There should be time for rchcarsals, inspections and debriefing.

The collection of war experiences provided an interesting example of good preparation for a night hunt by the 56th Army in the Southern Front during the Taganrog Operation (March 1942). A reconnaissance-in-force conducted several times proved unsuccessful throughout the Front sector. During preparations for the offensive, night hunts and ambushes in the sector of Riazhenaia and Solyeni heights were carried out for the purpose of capturing a "tongue." Although the raids did not provide results, the commander's actions in the preparations for the patrols displayed exemplary methods and procedures.[15]

The commander personally observed the area of the forthcoming operation, studied the nature of the German defense, and determined routes and orientation points. He assigned missions to the groups and individually to each soldier. All the patrol members were trained in command and control hand and arm signals, studied communication signals, set up passwords, and designated deputies. The group also worked through variants to German retaliatory activity while the scouts executed their mission. The commander finally checked each patrol member's understanding of the unit mission.[16]

How deep the "hunters" penetrated the German positions depended on the distance to the target and the nature of the terrain. As a rule, patrols usually did not exceed 10-15 kilometers total distance and usually penetrated the front line 3-6 kilometers in depth. To properly set-up and position itself, an experienced hunter team could remain in the German rear area for several days. The duration of the operation depended on how much was known about the target, how far it was from the front line, and how prepared the team was for activity in the German rear.[17]

For activities behind German lines, scouts had to be assigned missions with a specific target. The target required a specific purpose, and feasible obtainment. Scouts, or hunters, as a rule, were not to be diverted to missions for killing German groups which were encountered during the course of the operation. This was the mission of ambush patrols. The hunters sought to avoid and bypass encounters with Germans, since such diversions often gave an excuse to hunters for not capturing a tongue, which was their primary mission.[18]

Although the war experience study recommended that hunters receive their mission personally from the division commander or chief of staff, examples from Soviet writings indicate that the division and army chiefs of intelligence, except in special situations, handled operations and mission assignments.

Hunters were armed with automatic weapons, hand grenades, and knives. Traveling as lightly as possible for stealth and ease of movement, they would carry, depending on the situation, a gag, rope, and a flare pistol. They received fortified rations with high caloric content. Their equipment and clothing depended on the season and weather. Routinely, the clothing and equipment of the hunter teams resulted from the situation, the availability and types of materials. In some cases, the patrol members wore camouflage smocks, however, improvisation was the rule.[19]

For example, in late summer 1944, a hunter team outfitted itself from local and improvised materials. Their clothes were individually tailored for each scout from old rainproof tents. From a unit unique pattern, a seamster made a short collarless jacket and sleeves elbow length, with two false pockets and well fitted trousers to be worn over high boots. Instead of the standard issue boots each soldier was provided with shoes fashioned from double layered tarpaulin slippers on soft leather soles. In all the scout traveled very lightly with uniform, weapon, ammunition, tubes for breathing while immersed in the water, pliers, signal flare, gag, and smoke boxes. Ideally, the total weight carried by the scout was not to exceed 4 kilograms.[20]

The character traits of the personnel chosen for these missions were enumerated in Red Army General Staff study of war experiences and were a specific point of mention in other writings. Hunts were carried out by experienced scouts who possessed strong, steady nerves and were quick-witted. These bold individuals had to be capable of adapting themselves to the situation. Bravery, endurance, keenness, exceptional personal discipline, and a sense of small unit cohesion were primary qualities of scouts selected to participate in hunts.[21]

Colonel Voloshin, 39th Army chief of intelligence, noted, ". . . besides physical strength and dexterity, without which one is afraid of hand-to-hand fighting, at the moment of capturing a 'tongue' one must still have bold spirit, resourcefulness, and self-control. These qualities were ordinarily possessed by women. And, although this can seem at first strange, they served in our reconnaissance units."[22] He recalled being approached by some ten young women with a request, and at times, a demand to enter in reconnaissance. Many had been nurses, communicators, or in antiaircraft artillery. He remembered one young girl named, Maria Boiko, commander of a reconnaissance platoon in the 5th Guards Rifle Corps, who planned and led a hunt. During the patrol, a German soldier surprised them with submachinegun fire. Two of the hunters received wounds. After a short fire fight, four Germans were dead and a fifth wounded in the shoulder. Grabbing the wounded German soldier they called in artillery and mortar fire. Guards Lieutenant Boika with a flare pistol fired a red star cluster into the sky to mark their position. This drew fire from both sides.

To withdraw with wounded hunters and the prisoner was impossible. The group took cover. They fought off attempts by Germans soldiers to take their position. With nightfall and under the cover of darkness, Lieutenant Boiko led her patrol back to friendly lines. The wounded hunters were sent to the hospital; the prisoner, to the intelligence staff section. For her brave leadership in the battle, Guards Lieutenant Maria Boiko was awarded the Order of the Patriotic War, 1st Degree.[23]

When the front line traces took on a static disposition, the requirement for "tongues" increased in preparation for impending offensives. The Red Army command's penchant for detailed information was extremely high. As one post war analysis on reconnaissance concluded, ". . . during the last war it was sufficient to locate 60-65% of the enemy's locations to successfully break through defenses."[24] Such a comprehensive detail required almost seemingly inside knowledge. Consequently, on the static fronts, usually occupied by the rifle armies the hunt for tongues became a continuous and necessary mission. As one chief of intelligence summed, "The capture of 'tongues,' especially in periods when combat actions took on a positional character, became a highly difficult and dangerous affair. Hunters not infrequently paid for success with their blood."[25]

Besides the rifle units, armored units had a need for information prisoners. A chief of intelligence for a mechanized corps, in the early period of the war, recalled alternating hunts with ambushes and dispatching teams in "exhausting hunts for tongues."[26] As armored units in transition to an exploitation role from rear area locations in army and front operations in the latter stages of the war, their opportunities to patrol for tongues were limited -unlike the rifle armies which were always in contact along the front line.

Whether the rifle army was attacking or defending, hunter teams could be created. Close, broken terrain, such as forested, swampy terrain in the summer offered the best locations to conduct their "linguistics." During the winter months, patrols could be dispatched on skis over open terrain into German unit rear areas.[27]

Against fortified German positions, the hunt for tongues had to be covered by a diversity of reconnaissance methods. Voloshin gave an example of mixing patrols, "Today - night hunt, tomorrow - raid, day after - daylight hunt."[28] The diversity of methods indicates a repertoire of techniques within the hunts to keep operations from becoming stereotyped and predictable. The Soviet's diverse, active measures inhibited German plans to conduct counter-reconnaissance activities.

There were occasions when the daylight hunt was quite effective. Voloshin gives an example from his army. In preparation for the upcoming Belorussian operation, the 262nd Rifle Division in May, 1944, opposed German infantry forces near Vitebsk. The division intelligence officer notified Voloshin that they were thinking about a day hunt. The division's observations on

German positions opposite its sector established a pattern in that each night the Germans put out combat security to 150 meters east of the village Osinovka. In the morning the Germans withdrew to rest in a dug-out, leaving near the spot only two posts with sentries armed with submachineguns. Observers also noted that a small ditch was 60 meters from the German security trench. From this information a day hunt plan was worked out.[29]

In June 1944, in preparation for the Belorussian offensive, the various reconnaissance units within the 39th Army had, over the last twenty days before the offensive, conducted nearly forty hunts and captured sixteen prisoners, many with important documents.[30]

The information gained from "tongues" usually generated useful, tactical level intelligence. The information seldom possessed broad, long-range or background value. Instead, it usually applied directly to the immediate tactical situation. Most front line prisoners knew only a small piece of a larger puzzle. The movement of an army, the plans of higher commanders, the maneuver schemes of adjacent units was beyond their access. They know only what touched them. They knew their units, their fighting positions, their commander, their combat experiences, their weapons, their route to the front. They knew their routine necessary for survival at the sharp end of combat.

Sometimes "tongues" offered information that satisfied a commander's concerns. As General D. D. Lelyushenko, then commander of the 30th Army, recalled in his memoirs about his impending counteroffensive in December 1941. He directed the capture of "tongues." By the next morning six prisoners had been taken to include a small unit commander. Lelyushenko went himself to interrogate the prisoners. He had specific questions in mind. He determined that they did not suspect the preparation for the upcoming counteroffensive. "About the approach of our replacements and reinforcements, the prisoners knew nothing."[31] A commander gained confidence in his plan knowing he could achieve an element of surprise on the unsuspecting enemy.

The value of snatched "tongues" was most evident in immediate interrogation close to the battlefield. The stunned prisoner, shocked by capture, separated from his compatriots and friends, and cut-off from home, became vulnerable. No sophisticated interrogation techniques were necessary at the front - fear did the work. The fanaticism in the fight at the front between the almost religious orders of Nazism and communism led many Germans to believe in a natural Russian cruelty. This fear often resulted in an early cooperativeness by prisoners.

General Lieutenant A. N. Zaitsev, then a lieutenant commanding a regimental reconnaissance platoon, recalled a time when obtaining a "tongue" from the depth of the enemy defense became almost impossible. He remembered ravines and draws in which the slightest stir drew the attention of Germans who controlled the area and covered it with fire. "We were located with the enemy constantly in very close contact. The neutral sector ceased to be neutral. Very often it narrowed to not less than 60 meters that could be watched from both sides. In such a sector a grass snake could not crawl."[32]

In organizing a hunt patrol, the sector of patrol itself was carefully studied. The most convenient approaches and entry points were selected, and observation positions for observers and artillery observers fixed. Routes of movement of Germans patrols and sentry points were identified. Observation was conducted over the course of several days by the commander and part of the soldiers who were to participate in the patrol.[33] Hunt patrols often obtained documents listing the German units along the front line. Usually for the Red Army chiefs of intelligence, these just raised more questions like the percentage of strength for the listed unit. A priority element of information was the size and location of reserves. Additional information requirements raised by hunt patrols could be determined through further reconnaissance patrols or a reconnaissance in force.[34]

A captured German soldier with his soldier's book possessed a significant amount of useful and reliable information. In the soldier's book Red Army interrogators found the soldier's battalion and regimental unit numbers.[35] Pocket litter and other papers found on a captured "tongue" regularly betrayed interesting bits of information. Soviet intelligence analysts poring over uninteresting letters and seemingly unintelligible scraps of waste paper determined many useful facts about German units.

An anecdote related by the then assistant chief of intelligence for the 33rd Army, Major A.M. Sobolev, recounts an incident in November 1941. From one of the "hunt" patrols the scouts had captured the diary of a soldier from the 15th Infantry Division in fighting around Naro-Fominsk. The Soviets had fought desperately to hold the defense against the onslaught of the Germans and particularly the XXth Army Corps in this sector. In the diary the soldier had written:

"13 November. The Russians shoot all day long, and there is continuous night battle. From direct hits in the trench, shells have killed 7 and wounded 10 men. We all became very terrified.

14 November. Heavy artillery fire day and night.

15 November. Strong artillery fire. Three wounded.

16 November. At 1700 hours a direct hit on the next trench. Killed 5 and

wounded 2 men.

17 November. Terribly cold, many frost-bitten hands and legs. One man wounded."[36]

"These writings testified . . . that the Fascist soldier's blitzkrieg spirit was shaken," concluded Colonel Sobolev in his post-war account.[37] This was very important information as the Red Army scraped to hold their defenses on approaches to Moscow in the grueling winter fighting of 1941.

Intelligence analysis based on subtle, seemingly unrelated, bits of collected information from "tongues" shaped tactical decisions. For example, in September 1943, the 39th Army chief of intelligence received a summons to his new army commander, General Ivan Il'ich Lyudnikov. As the chief of intelligence recalled in his memoirs, "I reported to the commander about the enemy grouping, its strength and capabilities. I noted, from reconnaissance information, the weakest aspect appeared to be the 163rd regiment, 52nd Infantry Division."

"On what do you base such conclusions?" the Army commander asked.

"The regiment is filled with young soldiers, which according to a prisoner, will scatter with the first artillery strike. This information has been confirmed by visual observation," the chief of intelligence replied.[38]

Based on the chief of intelligence's conclusions, the army commander focused his main attack against the vulnerable sector. The attack advanced so rapidly it surprised defending Germans and paved the way for the recapture of Smolensk.

In January 1944, in the 39th Army sector, a lull in the offensive occurred. Aerial reconnaissance identified the approach toward the front of a column of German infantry. Their appearance signalled a potential change in the correlation of forces as calculated by the Army chief of intelligence. It became a collection requirement to find out the purpose of the troop movements. Were these units the next relief elements or attempts to reinforce the opposing unit?[39]

Colonel Maksim Afanas'evich Voloshin drove to the 91st Guards Division and shared his concerns with the division chief of staff and chief of intelligence. The division organized a hunter group to get a "tongue" to answer the important question.[40] The answer revealed the Germans were reinforcing the sector.

By this stage of the war many of the hunt leaders were seasoned veterans. In April 1944, Colonel Voloshin drove to the reconnaissance company of the 158th Rifle Division to assist in planning of a hunt mission. He met the leader of the planned hunter group, a senior sergeant named Mikhail

Yaglinskii. In two years as a scout, Yaglinskii had captured nearly thirty "tongues" with his patrols. He had been wounded five times, and after recovery each time had returned to the reconnaissance company.[41] This kind of experienced leadership provided extremely effective results.

The completeness and accuracy of the Soviet's intelligence on the German defensive dispositions proved very disconcerting to German commanders when they discovered the degree to which their dispositions had been discovered. The direct result of exposed defensive positions was Red Army operations which focused attacks against the weakness of German dispositions and plans.

In June 1944, during the Belorussian operation, the reconnaissance element of the 164th Rifle Division in Voloshin's 39th Army, captured two very important prisoners. They captured the 53rd German Army Corps commander, General Gollwitzer, and his chief of staff, Colonel Schmidt, who provided an interesting substantiation of the army's reconnaissance efforts. In candid conversation with General Ivan Il'ich Lyudnikov, commander of the 39th Army, Gollwitzer admitted, "You located our weak positions. Russian forces broke both legs, on which our corps stood. I did not understand from where you could have had such detailed information about our units . . ."[42]

Lyudnikov showed Gollwitzer the army's situation map, a compilation of Red Army reconnaissance efforts in the preparation period for the operation. On the map were markings of the 53rd Army Corps' units and its neighbors, system of defense and location of firing points. Voloshin described the German commander's reaction, "Gollwitzer for a while attentively studied the map. Then, looking at Schmidt pensively uttered, "If the inscriptions here were in German, I would consider that this was the working map which I used for the beginning battle."[43]

Marshal Rodian Yakovlovich Malinovskii, commander of the 2nd Ukrainian Front, in August 1944, had a similar testimony from captured enemy officers after the Yassy-Kishinev operation. He recalled in his post war account of the operation, ". . . that captured German and Rumanian staff officers admitted when shown maps of their defensive position that they believed them to be copies of maps made by their own hands . . . One could not be more delighted with the activities of our reconnaissance scouts."[44]

In the course of the Belorussian operation at the end of July, the 51st Army, 1st Baltic Front, advanced towards the Lielupe River south of Riga. Attempts to cross the river from the march had failed, and the army transitioned to defense. The 279th Rifle Division received the order to defend a 14.5 kilometer front along the river. Successfully fulfilling the assigned mission and preparing for subsequent offensive required exact information about German defensive positions, organization of battle groups, and fighting efficiency of its forces. This information gathered from different methods, included observation, eavesdropping, captured documents, questioning deserters and local inhabitants. Additionally, great attention was given to capturing information prisoners.

Natural and man-made obstacles inhibited penetration of the German sector. Separating the Soviet defense from the German positions the Lielupe river presented a wide water barrier approximately 100 meters wide with a depth of approximately 5 meters. The difficulty of crossing the water barrier was complicated by German engineer preparations. Their engineers had prepared obstacles along the banks with the use of concertina and barbed wire.

Difficulty in penetrating the German sector led the division chief of intelligence to direct that "hunter" teams be reduced to only two men. Preparation for capture of prisoners required a detailed plan which was worked out during the course of two weeks. From the 378th Reconnaissance Company twenty-four pairs of scouts were chosen. In the course of their preparation, the soldiers learned skills to combine stealth with resoluteness. Great attention was given to working out techniques to track down and capture prisoners. Experienced scouts who had captured "tongues" were the instructors.

The first three teams conducted hunts on the moonless night of 15 August. Each team operated independently, but against a single objective: "under the cover of night move to the assigned areas, discover their objective and without delay capture a prisoner."[45]

The first two teams were unsuccessful. One team had moved to their area and unexpectedly ran upon a German sentry. They struck him over the head, and he silently fell to the ground. Quickly the scouts shoved a gag in his mouth. Picking him up the hunters carried him to the river, where they discovered he was dead and had no documents. The second team had a similar experience, their prisoner died enroute to the river.

The third team, however, brought back a junior officer of the 1st Battalion, 272nd Infantry Regiment, 93rd Infantry division. He was immediately taken to the 51st Army staff for interrogation.[46] Information from one prisoner was insufficient and not desirable for analytic purposes. Intelligence analysts wanted corroborating information. However, to immediately hunt for "tongues" in an area where hunters had already visited was a dangerous risk.

The next team led by Sergeant V. K. Panov received the order to capture a prisoner in the area of farmstead Ozoly. They were prepared for the operation in this area. They moved into the area and found a spot in a garden 20 meters from a house and began to watch closely. They did not have to wait long. A German soldier came from the house. The scouts silently jumped on him, stuffed a gag in his mouth, tied his hands, and moved for the woods. Upon reaching the bank of the river, they placed the prisoner in a cork belt, and the scouts swam, pulling the buoyant prisoner, to the Red Army held bank. Both scouts received the order of the Red Star for this captive.

On 20 August, a larger group of five to seven men advanced at midnight to their designated area. On the northern edge of the village of Deni, members of the hunt patrol carefully opened the door to a dugout, and saw Germans sitting at the table. While one dozed, the second, a sergeant, was reading. Others in the room slept on plank beds. Carefully advancing, the hunters simultaneously struck the two seated soldiers. Shoving in gags and tying the prisoners' hands, the Soviet hunters began to drag them from the dugout. The sergeant, the bigger prize, was moved first followed by the other as they were pushed through the door. The hunt leader returned to the dugout door and with his automatic weapon killed the remaining Germans on the plank beds.

The return trip became difficult. The team fired red flares as a signal to artillery observers to initiate artillery fire to cover their recrossing of the river. At 0800 hours on 21 August, the prisoners, a sergeant and a corporal from the brigade "Gross," of the SS Nordland Division were delivered to the division staff and then to the army staff.[47]

For the period, from 10 August to 30 September 1944, division hunters captured seventeen tongues: one officer, two sergeants, two corporals, one lance corporal, and eleven soldiers.[48]

By late in the war, experienced hunters became very bold in their search for tongues. General Colonel Aleksei Zaitsev, a captain at the time, recalled in his memoirs a capture in late August 1944. He was commander of the 38th Rifle Division's reconnaissance group. During the Yassy-Kishinev operation, he received the mission to capture a Rumanian division commander. The Soviet High Command was interested in the Rumanian military leadership's reaction to Soviet radio announcements offering to accept the Rumanians who gave up resisting the Red Army.

Zaitsev led his team right into the division headquarters and the division commander's office. Pulling a pistol, Zaitsev told the Rumanian commander that the Soviet command wanted to talk with him. Zaitsev guaranteed his safety. Cautioning the Rumanian general to remain silent and to collect the papers on his desk, they walked out of the headquarters and got into the division commander's car. They drove off to the edge of town where they dismounted and began the trek back to Red Army lines. For Zaitsev, "unfortunately, I never found out about the fate of our biggest 'tongue' in all the war."[49] He never heard what became of his prisoner once he departed for higher headquarters.

The Rumanian division commander served a much greater purpose than an ordinary "tongue." Ultimately, however, the Soviet command could not reach any effective arrangements between the Soviet command and Rumanian authorities that would have led the Rumanians to abandon the German side of the war. Seizure of the Rumanian division commander may have been an attempt to secure a senior officer who would cooperate in making announcements over Soviet radio. Being forced to fight Rumanians and Germans in Rumania cost the Red Army 46,783 dead, 171,426 wounded, 2,000 guns and mortars, over 2,200 tanks and 528 aircraft.[50] The stakes were worth the risk of a hunt patrol.

In a recent publication for small unit commanders and military students, the organization and conduct of hunt patrols remains a basic requirement in tactical reconnaissance patrols. The hunt patrol is a basic method for the capture of prisoners, documents, and sample weapons.[51] The discourse specifies the lessons learned from past combat experiences and identifies the following major steps to conduct a hunt:

- Preparation of the unit for the hunt.
- Selection and study of the objective for the hunt, movement routes to and from the objective.
- Stealth movement towards the hunt objective and attacking it. The capture of prisoners, documents and sample weapons.
- Withdrawal to friendly force location.[52]

Reconnaissance played a significant role in the preparation of Red Army operations. Today, the information requirements for the Soviet Army's quantitative approach to warfare has increased the appetite of commanders and their staffs for more detailed information on enemy dispositions, the quantity of forces for a correlation of forces, and identification of enemy intentions. While modern intelligence collection systems working in the different disciplines of signals, imagery, and human resources has proliferated the battlefield, the Soviet Army has continued to make a significant investment in the use of tactical reconnaissance patrols. As their experience in the hunt for "tongues" shows, the Soviet Army is quite willing to pay in blood for combat information.

NOTES

1. M. A. Voloshin, *Razvedchiki vsegda vperedi,* (Moscow: Voenizdat, 1977), p. 17.
2. V. Chikin, "Razvedka v operatsiyakh 61-i armii," *Voenno-istoricheskii zhurnal* No. 10-1979, (Hereafter *Vizh*), p. 56.
3. This descriptive term for an information prisoner may have been used in Russian and Polish military lexicon since the Napoleonic era. During World War II, the term clearly meant an enemy soldier who would serve as a source of current information.
4. Voloshin, p. 18.
5. *Sbornik materialov po izucheniyu opyta voiny No. 2,* Sentyabr' -Oktyabr' 1942, (Hereafter Experience No. 2), (Moscow: Voenizdat, 1942), pp. 45-60; See *The Journal of Soviet Military Studies,* Vol. 1, No. 1, pp. 133-144, for example on the Soviet use of war experiences.
6. Ibid., p. 47.
7. Ibid.
8. U.S. Department of Army Pamphlet No. 20-269, *Small Unit Actions During the German Campaign in Russia,* (Washington D.C.: Department of the Army, 1953), pp. 208-213.
9. Ibid., p. 210.
10. Experience No. 2, p. 47.
11. Ibid., p. 51.
12. Ibid., p. 55.
13. Ibid., p. 54.
14. Ibid., p. 52.
15. Ibid., p. 53.
16. Ibid.
17. Ibid., p. 54.
18. Ibid.
19. Ibid., pp. 54-55.
20. A. Sivankov, "Za 'Yazykom' cherez lielupe," *Vizh* No. 9-74, p. 50.
21. Experience No. 2, p. 55.
22. Voloshin, p. 134.
23. Ibid., pp. 136-7.

24. F. Gedrasov, *Voenii Vestnik,* No. 3-1987, p. 3.
25. Voloshin, p. 18.
26. D. A. Dragunskii, *Gody v brone,* (Moscow: Voenizdat, 1975), p. 87.
27. Experience No. 2, p. 54.
28. Voloshin, p. 128.
29. Ibid., pp. 132-33.
30. Ibid., p. 154.
31. D. D. Lelyushenko, *Moskva-Stalingrad-Berlin-Praga,* (Moscow: Nauka, 1973), p. 98.
32. A. N. Zaitsev, *Na ostrie krasnykh strel,* (Mosocw: Voenizdat, 1988), pp. 59-60.
33. Experience No. 2, p. 54.
34. Voloshin, p. 70.
35. Ibid., p. 17.
36. A. M. Sobolev, "Boi pod Naro-Fominskom," *Bitva za Moskvu,* (Moscow: Moskovskii rabochii, 1985), p. 254. Sobolev later in the war became the chief of intelligence for 1st Guards Tank Army.
37. Ibid.
38. Voloshin, pp. 79-80.
39. Ibid., p. 110.
40. Ibid.
41. Ibid., p. 120.
42. Ibid., p. 171.
43. Ibid.
44. R. Ya. Malinovskii, *Yassko-Kishinevskie Kanny,* (Moscow: Nauka, 1964), p. 55.
45. Vizh 9-74, p. 51.
46. Ibid.
47. Ibid., p. 53.
48. Ibid.
49. Zaitsev, p. 173.
50. John Erickson, *The Road To Berlin,* (Boulder, Colorado: Westview Press, 1983), p. 371.
51. F.I. Gredasov, *Podrazdeleniya v rasvedke,* (Moscow: Voenizdat, 1988), p. 201.
52. Ibid., pp. 205-209.

CHAPTER 6

Red Army Battlefield Agility: The Legacy

Modern battlefield lethality spans ever greater widths, depths, and heights at distances a unit commander can no longer observe from a single vantage point. On this spacious battlefield, the real correlation of opposing forces is determined by actual fighting capabilities which are brought to bear. Victory requires more than overall power; it necessitates superiority of forces at key points to inflict losses and defeats.

Armored forces remain the principal means of satisfying modern imperatives for highly mobile fighting at chosen times and places on the battlefield. Through agile maneuver, they can deliver soldiers and firepower quickly to create decisive combat advantage. This agility is increasingly necessary for survival, as well as victory.

We have stereotyped notions of how various national armored forces performed during the Second World War. We perceived some armies as having been agile in their maneuver of armored forces while we view others as only using tanks to provide support for the infantry. Armies which exploited the mobility of armored forces understood practical requirements for agile movement - an understanding forged in combat trials absent in peacetime exercise. Regardless of how any army used its armored forces, its experience contributes to a practical wisdom that shapes its contemporary armor doctrine, force structure, and combat expectations.

Pursuit of Red Army battlefield legacies clarifies our assumptions as to how the Red Army fought its armored forces and provides insights to their solutions for this complicated problem of mobile armored warfare today. Review of past Red Army experiences reveals answers to such questions as: How agile were Red Army armored forces? Did they achieve agility in any current sense? What expectations and capability do today's Soviet armored forces have to achieve battlefield agility? The practical legacy of Red Army armored forces provides a context for analyzing current and future Soviet performance. If agility is a mental as well as physical quality, as suggested by some military writings, then expectations become a very critical dimension of capability.

The U.S. Army AirLand Battle concept defines agility as "the ability of friendly forces to act faster than the enemy - [it] is the first prerequisite for seizing and holding the initiative."[1] Elements of this basic tenet imply that ". . . greater quickness permits the rapid concentration of friendly strength against enemy vulnerabilities."[2] Agility depends on the commander's ability to "read the battlefield," to decide quickly, and to act without hesitation. "Agility is as much a mental as a physical quality . . . formation at every level must be capable of shifting the main effort with minimum delay and with the least possible necessity for reconfiguration and coordination."[3]

While these doctrinal phrases are acceptable in writing, there is a lingering realization that theoretical pronouncements always defy easy practical application on chaotic battlefields. To explore mobile armored warfare is to delineate practical prerequisites that enable agile forces to successfully survive in continuous combat.

In the 1920s and 1930s, the revolutionary Red Army eagerly embraced the wartime potential of new technologies, like the combustible engine. Unlike western military theorists who advocated small professionally mechanized forces, Soviet military thinkers endorsed large scale mechanization and motorization. By 1933, the Red Army had pioneered large mechanized forces by forming two mechanized corps, five mechanized brigades, two separate tank regiments, twelve mechanized regiments, fifteen separate tank battalions.[4] A large-scale maneuver in the Kiev Military District conducted in 1935 proved to the Red Army leadership, if not to all foreign military observers, that mechanized units were a factor which radically altered the nature of combat.[5]

Large mechanized forces, however, complicated considerably both movement and combat. These complex modern engines of war proved to be just as insistent as the men and beasts of past wars; they required rest, repair, refueling and rearming. They consumed large quantities of fuel and required spare parts and maintenance beyond previous logistical requirements. Logistical sustainment became an increasingly integral part of decisions to move and employ armored forces. Command and control became stretched, necessitating precise efforts to mass, protect, and fight the force.

Simultaneous with practical armored force re-structuring, Red Army military theorists, such as V. K. Triandafillov, K. B. Kalinovskii working in concert with Marshal M. V. Tukhachevskii, developed a concept of deep operations using airplanes, airborne forces, and mechanized units to strike simultaneously the entire depth of the enemy's disposition. This ambitious concept, as yet beyond the immediate technical means and force capabilities of the Red Army, became its theoretical goal.

The Red Army's bold, early commitment to use of massed armored forces waned in the late 1930s as Stalin's purges eliminated many of the major theorists. Additionally, interpretation of the Spanish Civil War experience cast doubt on the feasibility of independent use of tank forces.[6] Top Red Army advisors from the Spanish Civil War saw tanks first as infantry support weapons with less reliance on large mechanized formations in initial operations. The Red Army's combat actions seemed to confirm shortcomings in mechanized corps. During the invasion of Poland in 1939, the Red Army's cumbersome tank units (two tank corps) experienced logistical problems on long movements. Limited signal communications proved woefully inadequate and complicated command and control.[7] Later during the Finnish War (1939-40), large armored units were of little value during winter operations in the extremely rugged and forested terrain.[8] Based on these and other experiences, by fall 1939, the Soviet leadership decided to dismantle the large mechanized forces and placed smaller tank units in direct infantry support creating more balanced motorized divisions each with 275 tanks.

At the same time, the success of General G. K. Zhukov's forces at Khalkhin-Gol and Soviet theorists' analysis of German armor successes against Poland and France in September 1939 and May 1940, respectively, prompted a reconsideration of large mechanized forces. A high level decision in June 1940 reinstated the mechanized corps (first nine and then twenty more). Their initial formation in spring 1941 ensured an immature mechanized force at the time of the German invasion in June 1941.

These huge formations consisted of two tank and one motorized divisions with 1,031 tanks. With too few radios, command and control of these units by inexperienced commanders who had survived the purges remained difficult at best, particularly at the tactical level. Their subsequent performance in 1941 revealed a number of problems masked by overly managed peacetime exercises, which widened the inevitable gap between the ideal and practical. This is a recurring lesson: an army should clearly understand that it cannot fight wars on theoretical musings. An army must think practically about its own operational capabilities.

By 22 June 1941, the forward Soviet military districts had deployed more than a dozen mechanized corps. However, the corps lacked a full complement of personnel and equipment.[9] While they possessed an impressive number of armored vehicles, many of these were outdated, poorly maintained light tanks, derisively referred to by Red Army tankers as "sparrow shooters" or "knights in plywood." Only a small percentage were the new KV and T-34 tanks. Artillery had inadequate ammunition, staffs were incomplete and untrained junior officers led platoons and companies. According to one Soviet observer, "the cumbersome mechanized corps in conditions of understength staffs with untrained personnel, along with their poor cohesiveness and insufficient number of radios proved difficult to control on the battlefield."[10] The Red Army had rebuilt the identical structure that presented problems in Poland.

The Red Army assigned the majority of its mechanized corps to rifle armies. When fighting began against the German invasion in June 1941, decentralized employment of these corps precluded focus of their combat power. Early in the frontier battles, the situation transcended brave Red Army soldiers or cowards, good commands or bad, planning or lack of it. All were in the grip of raging, uneven battles.

The mechanized corps fought after marching great distances under continuous German air attacks which resulted in debilitating losses from devastating air strikes. On arriving in combat positions the corps could neither rest their personnel nor perform maintenance on combat vehicles.[11] In the Northwest Front, the 3rd and 12th Mechanized Corps fought against the Germans advancing along the Shaulayai and Vilnius directions. Elements of the two mechanized corps counterattacked against the German main effort along the Tilsit-Shaulayai highway. Four German air strikes delayed the 12th Mechanized Corps attack to late morning. By afternoon, its attacks had engaged large German forces and were unsuccessful. The 3rd Mechanized Corps suffered severe losses from German air strikes as well.

By 24 June, the 3rd Mechanized Corps' 2nd Tank Division ran out of fuel and was destroyed. On 25-26 June, parts of the 12th Mechanized Corps moved from encirclement and retreated to the Western Dvina river. By the end of 25 June, the Northwest Front was in shambles. Charging like angry rhinoceroses, the Front's mechanized corps attacked frontally into the strength of the German spearhead. Erratic instructions from Army and Front commanders created eccentric commands that further dispersed the mechanized corps and dissipated their combat power.[12] German commanders, initially surprised by the ferocity of the Red Army assaults, adjusted quickly and destroyed their more clumsy antagonists.

The Western Front's 6th, 11th, 13th, and 14th Mechanized Corps suffered a similar fate. The 11th Mechanized Corps located too far forward at Grodno on the 22nd June, immediately became consumed in the advancing German wave. The 14th Mechanized Corps ordered to attack on 23 June, was too widely dispersed and could not bring its subordinate units together. The 22nd Tank Division dissolved in uncoordinated and unsuccessful meeting engagements and withdrew from the fight the morning of 24 June. The 13th Mechanized Corps, a corps in name only for it lacked its complement of tanks, was swept away and then encircled near Belostok.

In the Western Front sector, planned, coordinated mechanized corps counterattacks failed to materialize by 25 June, ". . . due to the almost total lack of air support and insufficient artillery, particularly antiaircraft artillery. The losses in men and equipment from enemy shelling and air pounding were great. Because of the Nazi's ceaseless air raids supplies and communications and fuel bogged down."[13] These problems were compounded by a collapse of command and control.

The Southwest Front included 4th, 8th, 9th, 15th, 16th, and 22nd Mechanized Corps. The 4th and 22nd Mechanized Corps spread widely in their initial dispositions conducted unorganized counterattacks across a wide front. Many of the mechanized corps were located too far from their assigned area of operations, requiring time to concentrate and long march routes exposed them to German air strikes. The 15th Mechanized Corps, the strongest of the corps with 133 KVs and T-34s, conducted an ineffective broad frontal attack. Some of its units ran into trouble with wooded-swampy terrain, and the old nemesis of poor command and control became evident. Lacking sufficient radio communications, corps staffs operated with only two of eight authorized radio stations.[14]

The Southwest Front's fight, however, endured longer than in the other front sectors because of the depth of mechanized corps dispositions. The 4th Mechanized Corps reinforced with two artillery regiments fought strongly at Lvov. The 8th and 15th Mechanized Corps counterattacked against the southern flank of the German advance on 26 June. After extensive marches, they failed to coordinate their attacks with 9th and 19th Mechanized Corps attacks northeast of Rovno. The 8th Mechanized Corps reported German aircraft using a gasoline mixture like present day napalm to destroy tanks. Southwest Front resistance prompted the Chief of the German General Staff, General F. Halder, to note in his diary on 26 June, "Army Group South is advancing slowly, unfortunately with considerable losses. The enemy on this front has energetic leadership."[15] And on 27 June, ". . . the Russian Command in the Ukraine . . . is doing a pretty good job . . ."[16]

Review of the collective experiences by mechanized corps in the frontier battles reveals a number of prerequisites for battlefield agility in armored operations. Fundamental is the ability to move through the environment. As lucrative targets, large mechanized formations needed protection from German air attacks. The Soviet Air Force, devastated on the ground by attacking German Luftwaffe in the opening hours of the invasion, simply was unavailable to protect ground forces. Only rudimentary air defense existed in all armies during the war. Mechanized corps, unable to move freely over Luftwaffe-dominated terrain, missed start times for attacks and suffered heavy, irretrievable losses.

Sustainment of the action requires logistics as well as focused command and control. Orders had to be directed towards specific missions or objectives to ensure concentric actions by the forces. In these early battles, commanders who thought they knew the situation and could find their corps tried to send them everywhere. After two days of fighting most mechanized corps sat like mortally wounded dinosaurs with neither fuel nor ammunition to continue the fight, and unable to move out of harm's way.

Poor mechanized corps performance became an important catalyst for change by providing early lessons to be acted upon by the Red Army if it was to survive the thrusts of German panzer forces. By fall of 1941, mechanized corps had been abolished and replaced by smaller, more manageable separate tank brigades and battalions. These truncated armored units were more easily controlled and more effectively fought at the tactical level.

The scarcity of new tank brigades required them to be agile on the battlefield, both in their tactical fights to survive and their operational movement among main sectors. The 4th Tank Brigade under Colonel Mikhail E. Katukov defending before Mtsensk in October 1941 provides a dramatic example which foreshadowed the slow rebirth in Red Army armored forces. As a survivor of the early armored battles in the Ukraine, Katukov, who commanded the 20th Tank Division, 9th Mechanized Corps in the Southwest Front's fighting at Rovno, applied lessons learned to compensate for the inexperience of his tank commanders and crews. He fought an active defense from ambush positions that required little movement over ground which the tankers knew. His reserves made direct line attacks to enhance their fighting effectiveness. His tactical deception of building dummy defensive positions disguised locations of his strongpoints from attacking German armor and misled German air strikes. Today, Soviet military analysts see a similar correlation. The number of attacking tanks which manage to penetrate an antitank defense decreases as the number of dummy positions increases.[17]

Tenacious, expert handling of the tank brigade inflicted serious casualties on the German panzer group and caused their overextended supply lines to snap. Katukov's fight brought Guderian's tanks to a crawl, achieving only 30 kilometers in eight days.

Katukov's performance provoked Guderian to write in his memoirs, ". . . the quality and, above all, of the new tactical handling of the Russian tanks were very worrying."[18] Guderian a little further in his account conceded, "The damage suffered by the Russians was considerably less than that to our own tanks."[19]

After eight days of continuous fighting, the 4th Tank Brigade withdrew from battle near Mtsensk to be employed to the north. Katukov elected to road march 300 kilometers instead of risking German air strikes while on the railroad. Katukov's brigade later fought defensive battles again on the German's northern approach to Moscow.

However, during the winter counteroffensive of 1941-42 around Moscow, the use of cavalry corps and small tank units failed to turn tactical gains into operational success, nor did it produce sustained operational tempo.[20] These experiences demonstrated the need for larger armored units designated for faster independent actions at greater depths. Larger tank units were required to envelop large encircled enemy groupings.

In 1942 increased tank production by displaced factories permitted formation of tanks corps by spring, and mechanized corps by fall.[21] Tank corps consisted of three tank and one motorized brigades (168 tanks). Mechanized corps had three mechanized and one tank brigades (175 tanks). Truck shortages limited the number of mechanized corps and during the summer the Soviets created two tank armies around the nucleus of tank corps.

These new armored formations were forged in a baptism of fire. The disastrous Kharkov operation in May 1942, squandered three of these new tank corps through inept operational handling, precluding any test of their tactical mettle. Subsequent mobile armored warfare in the southern sector of the Soviet-German front during the remainder of the year were fought with many mistakes - and lessons.

For example, in the Red Army General Staff's "Collection of Materials for the Study of War Experience," The 1st Tank Corps, on 20-21 November 1942, ". . . instead of going around the enemy and exploiting in depth engaged the enemy . . . [its] reconnaissance unable to discover the presence of the 22nd Panzer Division . . . [and] clash[ed] unexpected[ly]."[22] The study concluded, "timely reconnaissance would have precluded loss of time in changing march formation while moving for executing detours or for deploying parts of the force for attack."[23] Red Army analysis of combat experience recognized, "ground reconnaissance should work closely with air . . . [and] must have adequate signal communications equipment. Behind reconnaissance, detachments for making movement possible [were needed for] . . . removal of obstacles, marking detours, regulators of movement."[24]

In another study of Stalingrad and Don River operations, Soviet analysts noted a lack of "... favorable examples of cooperation between air army and mobile troops . . ." and concluded, it would be ". . . expedient to place ground attack aircraft in the operational depth under the command of the tank and mechanized corps."[25]

Conclusions derived from late 1942 experiences reinforced earlier requirements gained from the frontier battles: air cover and now engineer support were necessary to move through the environment. The Red Army also lacked dependable artillery support and close coordination with aviation for fire support to 'pave the way' for mobile troops. Secret regrouping was required to create surprise and attain operational success.[26]

Regroupings under a cloak of deception and operations security allowed forces to move uninhibited. It was increasingly necessary to see the battlefield and poor reconnaissance against the German's produced obvious results.[27]

Analysis of armored warfare experience during the first period of the war allowed the Red Army leadership to refine tactical and operational use of tank and mechanized formations. New principles were directed in the People's Defense Commissar (NKO) Order No. 325, dated 16 October 1942. According to the order, separate tank brigades and regiments were designated for direct infantry support. Tank and mechanized corps were Front and army means, to be used on main axes as an echelon to develop success by inflicting powerful strikes to split up and encircle large enemy groupings.[28] These generalized experiences, passed in the form of directives, prompted Red Army commanders to incorporate quickly lessons learned. Tank commanders of 1941 tank brigades led corps in 1942, and provided a core of hard, experienced survivors.

The first new tank armies, 3rd and 5th, formed in April-May 1942, were mixed structures, formed from tank corps, rifle divisions, cavalry divisions or corps, artillery and mortar units, and various support elements. At the time, Red Army thinking believed tank armies could independently penetrate prepared tactical defenses and develop tactical success into operational.

However, experience quickly proved such a structure too slow and awkward to do so. Tank army commanders were unable to manage their diverse forces and organize continuous cooperation. Varying capabilities of mobility, and a lack of effective communications, seriously hampered troop control during development of an offensive into the operational depth. With combat power weakened and the depth of tank strikes limited, operational success deteriorated rapidly. Weaknesses became evident during the Stalingrad operations. Red Army leadership was learning the connection between force structure and agility.

Operational experience showed the necessity for creating uniform, highly mobile armies which possessed greater strike and fire power. For tank armies to obtain decisive results required freedom from the slower, less agile non-motorized infantry formations, greater mobility for artillery and rear services support, and provision of the army with engineer support.[29]

The development of a homogeneous force structure produced more agile formations. In January 1943, the 5th Guards Tank Army, with two tank and one mechanized corps, appeared. Then, in rapid succession in April, 1st and 2nd Tank Armies were reconfigured to the new organization and, in May and July, the 3rd Guards and 4th Tank Armies, appeared. These armies consisted from one two-tank and one mechanized corps, a separate tank and self-propelled artillery brigades, artillery formations, reconnaissance, engineering, chemical and other support units. The use of brigades and corps configuration permitted greater flexibility and ultimately agility. It also allowed variation in the numbers and types of battalions, so that tailored forces such as the forward detachments could emerge. The combat strength of tank armies numbered 600-700 tanks and self-propelled assault guns, 500-600 guns and mortars, and 30-35,000 men.

In practice, the composition of every tank army varied and in relationship to the combat operation depending on such factors as situation, mission, and availability of forces. Such task organizing is a notion not normally ascribed to Red Army units, because this qualitative alignment of forces requires an understanding beyond numbers. Yet, these considerations were evident in the Red Army during World War II.

By summer 1943, the Soviet High Command had created five tank armies as mobile groups to operate on Front main axes of advance to develop success in the offensive and inflict counterstrokes in defense. In January 1944, the Red Army formed its last wartime tank army, 6th Tank Army.

While tank armies evolved, the tank corps under very capable leadership continued to demonstrate an agile base upon which tank army commanders would depend.

The 10th Tank Corps operations during critical battles on the southern face of the Kursk salient gives a good example of the 'mettle' of Red armored forces by mid-war. This typical corps, with a typical commander by the summer of 1943, like other corps, had accumulated a large reservoir of command and staff expertise. This example evidences that with agility one not only chooses time and place of engagement but also type. (see map)

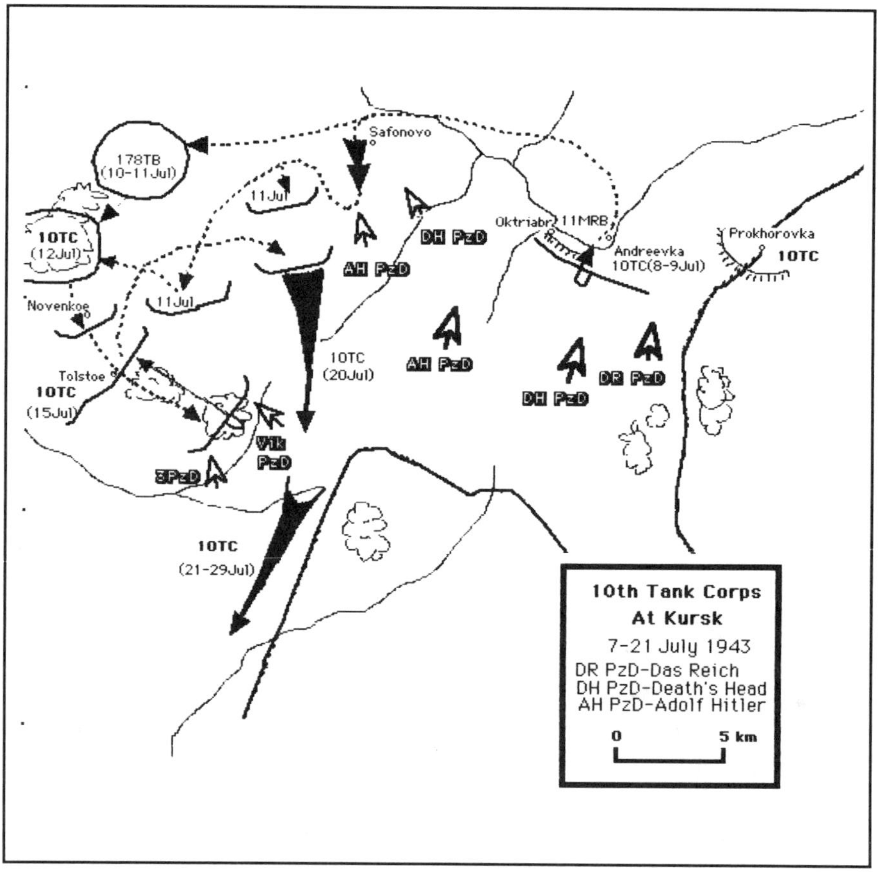

On the night of 6/7 July, Stavka assigned 10th Tank Corps to the Voronezh Front and directed its movement to the Prokhorovka area. At 0100 hours, 7 July, General N. F. Vatutin, Voronezh Front commander, ordered General Lieutenant V. G. Burkov's 10th Tank Corps to concentrate specifically in the Petrovka-Aleksandrovskii area by evening of 7 July to prepare a defense against an attack from the southwest, and to cut the road to Prokhorovka from Belenikhino.

The corps commander assigned missions to his brigades on the march. He prepared his brigades and regiments to conduct a counterstrike from the march in a meeting engagement upon arrival in the assigned area. Employing road guides, the tank corps moved by forced march across open terrain without fighter coverage above along two previously reconnoitered routes. Luckily the corps was not caught by marauding German warplanes.

By midday 7 July, the 183rd Tank Brigade's forward detachment arrived at Prokhorovka. Met by corps staff officers, the brigade was immediately directed to their defense area. At 1300 hours, the brigade assumed a defensive line south and southwest of Prokhorovka with the brigade's motorized rifle battalion south of Teterevino and a reconnaissance element to Malo Yablonov.

During the next four hours, the corps concentrated its remaining combat units south and southwest of Prokhorovka. Units took positions; soldiers and tankers entrenched themselves and camouflaged tanks. The commanders established a system of defensive fires, coordinated defensive actions, and established communications.

With the advance of the elite German II SS Panzer Corps to the area of Teterevino, Malye Mayachki, General Vatutin, the Front commander, decided to inflict a counterstrike against the advancing German armor group with 10th Tank Corps and other units to restore the situation. At 2300 hours, 7 July, he ordered by radio 10th Tank Corps to push forward to a line Krasnyi Oktiabr-Andreevka, and at 1000 hours, 8 July, strike together with the 2nd Tank Corps in the direction of Vasilevka, Yakovlevo, to destroy the Germans. Simultaneously, the corps was to prepare to repel any potential enemy strike.[30]

Early morning 8 July, the corps advanced to the assigned line and assumed previously prepared positions in the army rear sector. Combat vehicles were dug-in and carefully camouflaged from enemy observation and the corps placed the motorized infantry forward of the position for security and began reconnoitering the area.

Front forces delayed the German advance while the counterstroke was being prepared. At 0600 hours, 8 July, the German XXXXVIII Panzer Corps with aviation support attacked north. Elements of II SS Panzer Divisions, 'Adolf Hitler' and 'Das Reich,' struck 10th Tank Corps and, by 1300 hours SS Panzer Division, 'Totenkopf,' joined the attack.

The German attack was preceded by powerful air and artillery strikes. The 10th Tank Corps repelled the first attack, but the second effort allowed German forces to drive forward. A 10th Tank Corps counterattack with two brigades, although unsuccessful overall, did force the 'Totenkopf' Division to stop and regroup. Tenacious fighting all day forced the commitment of the 183rd Tank Brigade's reserve, a tank company. By the end of the day, 10th Tank Corps had repelled six German attacks.

Using the cover of darkness, the corps healed its wounds, executing intensive staff work, maintenance, resupply and medical care. During the night, the corps received the mission: leave on the front line the 11th Motorized Rifle (MR) and 173rd Tank Brigades, by morning of 9 July, the corps main force was to concentrate in the vicinity of Zorinskie Dvory, where it was to be subordinated to the 1st Tank Army.[31] The corps commander received his instruction from the 1st Tank Army commander, General Mikhail Katukov, and returned to his brigades. The two brigades' combat missions were received while they tracked along their march routes. At 1100 hours 9 July, the brigades with separate artillery regiments accomplished a lateral regrouping on the Oboyan direction, assembling near the village of Ship.

As the corps moved west, SS 'Adolf Hitler' and 'Totenkopf' Divisions, hit part of the 1st Tank Army in the vicinity of the Solontinka River line, and began to press them northwest. The situation became critical and created a worrisome threat of a large German armor penetration in the direction of Oboyan. General Katukov, fighting another serious defensive battle, gave the corps the following order: "Units of the 10th Tank Corps from the line Orlovka and Safonovo at once inflict a counterstrike against the penetrating enemy and restore the situation."[32]

From north of Melovoe the 183rd and 186th Tank Brigade (TB) and the 1450 Self-propelled (SP) Regiment and infantry riding on combat vehicles advanced aggressively in columns. The 10th Tank Corps, with nearly 100 tanks and SP artillery, deployed into combat formation initiating a meeting engagement. The momentum of the assault drove the Germans back to their initial starting positions. Holding on their limit of advance, the 186th repelled four German counterattacks; the 183rd, one.

In the afternoon, the 183rd, withdrawn from battle, marched southwest. By 2100 hours the tank brigade accompanied by the 727th Assault Gun Regiment assumed a defensive line Kruglik, Kalinovka.

Early morning on 10 July, Red Army reconnaissance reported that German regrouping for a resumed attack against the 1st Tank Army was imminent. The 186th Tank Brigade and 1450th SP Artillery Regiment moved southwest. At 0300 hours they assumed a defensive position at Kalinovka, extending their right flank to the 183rd Tank Brigade.

On the morning of 10 July, the tank corps' remaining tank brigade, 178th, became the corps reserve in the area of Kursovsk. The corps commander and his staff reconnoitered and planned potential counterattack routes for the reserve. Terrain before the corps defensive position was a broad sloping plain to the south, cut by ravines, small rivers and streams. The landscape was dotted with villages and farms and groves - ideal tank fighting country. Here for the next five days the armored battle would rumble on.

One corps veteran described the actions during this period, "... day was continuous and bloody battle, night was exhausting marches, and then dig trenches and cover for the tanks. So shaped the actions that saved us from losses: in battle we saved people and combat equipment."[33]

At 0400 hours 10 July, in the corps defensive sector a powerful artillery and air strike preceded the German 11th Panzer and parts of the GrossDeutchland Divisions' attack. In heavy fighting, the corps stood its ground well into the night. However, by evening the 1st Tank Army situation on the right flank had seriously deteriorated. Penetrating the tank army's 6th Tank Corps the German units actively sought to encircle and destroy part of the tank corps. In the early morning hours on 11 July, a group of tanks with machine-gunners riding on top from the 183rd Tank Brigade was detailed to assist the beleaguered tank corps. They attacked penetrating German units in the flank, stopping any further advance. Avoiding encirclement, the 6th Tank Corps withdrew to the area of Noven'koe. The same morning 10th Tank Corps' separated 11th Motorized Rifle Brigade continued to defend along a line Krasnyi Oktiabr, Prelestnoe, against heavy pressure.

The II SS Panzer Corps regrouped during the night and most of the morning of 11 July. Throughout the 11th, the 11th Motorized Rifle Brigade fought to hold the line at Krasnyi Otiabr. The same day, the German XXXXVIII Panzer Corps continued to attack in the Oboyan direction. The tank brigades held against repeated attacks. During fighting against the 183rd and 186th Tank Brigades, the corps commander had to commit his reserve of a guards rockets battalion and a battalion of assault guns. "The timeliness and correctness of this decision determined the course of further fighting," a unit historian noted.[34] The situation had been stabilized and the line held. From 7 to 11 July, the corps continued to conduct defensive battles on the approaches to Oboyan and Prokhorovka playing a significant role in the Red Army defense at Kursk.

Again, in the early morning hours 12 July, the 10th Tank Corps received orders from General Katukov, to turn over its sector to a rifle division and conduct a forced march, and by 0700 hours 12 July, concentrate in the woods three kilometers northwest of Noven'koe. The corps maintained its strength remarkably at 100 tanks and assault guns. The corps staff planned and coordinated an attack with 5th Guards Tank Corps against the flank of the II SS Panzer Corps advancing in the direction of Prokhorovka. Under the cover of darkness, corps sappers assumed positions on the swampy banks of a small river, and made three crossings in a few hours. By early morning, the corps commander checked brigades' and battalions' last minute preparations and instructed them to operate aggressively, sharply, and decisively.[35]

At 0800 hours, 12 July, after a powerful artillery and air preparation, the corps advanced. In the first echelon attacked the 183rd and 178th Tank Brigades reinforced with assault guns and a mortar regiment. The 186th Tank Brigade and SP artillery regiment followed in the second echelon. By 1700 hours, the brigades had advanced 12-15 kilometers. The 178th entered battle for Berezovka; 183rd, for Verkhopen'e; and 186th, for Tolstoe. Fighting well into the night, the Germans conducted strong counterattacks in the vicinity of Verkhopen'e and Berezovka. Combined with 5th Guards Tank Army and other units, the 10th Tank Corps fought around the periphery of the greatest tank battle in military history.

On 13 July, German forces with the support of numerous ground attack aircraft, attempted to knock out the tank corps' units in the vicinity of Tolstoe, but suffered heavy losses and withdrew under pressure. After seven days of battle, the 10th Tank Corps operated with 86 tanks (56 T-34 and 30 T-70).[36]The corps' unit history gives rightfully great credit to the rear services and maintenance for this sustained feat.[37] Red Army tank units now held the battlefield. This meant they could recover and restore damaged tanks. A combination of fearless recovery under fire and dedicated long hours in maintenance returned a significant number of combat vehicles. Some German commanders, when leaving the battlefield, ordered no damaged Soviet tanks be left salvageable because of the Red Army's combat vehicle restoration ability.

Regrouping its forces and equipment, the German command resumed their fight on the morning of 14 July. By the middle of the day, the 3rd Panzer Division broke through a forward rifle division and found the boundary between the 5th Guards and 10th Tank Corps, nearly overrunning the command post of the 178th Tank Brigade. The situation on the corps' left flank continued to worsen. German panzer units pressed the adjacent brigades, and took ground near Tolstoe, advancing to the rear of the 183rd and 186th Tank Brigades while attacking them from the front. By afternoon the German ground advance, accompanied by powerful and devastating air attacks, had created a ferocious battle. Again, close and heavy fighting raged until late night. The 10th Tank Corps held its ground while the Germans spent their force. By evening, panzer units too weak to continue their advance, paused frustrated and blown.

At this time, the Voronezh Front commander resubordinated 10th Tank Corps to 6th Guards Army. Ordered to withdraw, the corps moved to a more advantageous defensive line along height 240.2, southern edge of Noven'koe, and height 247.5. Occupying these positions by afternoon on 15 July, the corps exchanged desultory fire with German units on 15 and 16 July.

While occupying the defensive line, the corps began preparation for attack. By 16 July, the corps had fifty-eight tanks and assault guns but only eighteen T-34s with crew were immediately available for reinforcement. These were sent to the 178th and 183rd Tank Brigades.[38] By the end of 19 July, restored tanks and additional crews increased another seventeen tanks for a total of ninety-three tanks with crew.

On the morning of 20 July, the 10th Tank Corps became resubordinated to 5th Army, and received orders to attack in the direction of Novoselovka. Again, the corps' combat formation advanced in two echelons: in the first echelon 186th and 183rd Tank Brigades reinforced by self-propelled artillery and assault guns, and in the second echelon the 11th Motorized Rifle Brigade with a mortar regiment. The 178th Tank Brigade reinforced with an assault gun regiment became the corps reserve. At 13 hours, after a strong artillery strike along Novoselovka and Sukho-Solotino, 186th and 183rd Tank Brigades began the corps attack along the Oboyan-Belgorod highway.

Over a relatively short time, 10th Tank Corps operations consisted of a variety of engagements: defense, meeting battle, attack, delay. Accenting the ability to sustain operations, the tank force demonstrated a mobility and agility that allowed Red Army commanders not only choice in time and place for its fights but also type of engagement.

While a tank corps could demonstrate this kind of agility, a similar performance and effectiveness by tank armies remained measured and costly in execution. Red Army tank army commanders still had much to learn in operational art before tank armies would demonstrate an agility comparable to the 10th Tank Corps tactical actions.

But, practical experience was building. From their *Collection of Materials for the Study of War Experiences* the following was noted:

> The movement of tank (mechanized) corps over great distances by road must be avoided. This method of assembling troops may be resorted to

> in exceptional cases only. Long road marches preceding an operation lead, as a rule, to a great expenditure of motor fuel, wear and tear of equipment, fatigue of personnel and fall out of tanks due to mechanical defects, particularly when going through roadless country. This can be confirmed by taking as an example the concentration maneuver of the 2nd Tank Army, in February 1943. This army had to complete a march through roadless country over a distance of more than 200 kilometers and lost about 55% of its equipment.[39]

After Kursk, the Red Army conducted a counteroffensive, using its tank armies to race the German Army for the Dnepr River. During the advance, employment of composite detachments and the use of ad hoc formations and structures within some tank armies reveals an interesting dimension in agility. In this operation, as in later instances, the combat situation did not allow tank units, in spite of large losses, to be withdrawn from fighting for restoration and reformation of their combat strength. With relatively low strength in manning and combat vehicles, composite detachments had to be formed up.

In August 1943, while driving the Germans westward towards the Dnepr River, the 5th Guards Tank Army possessed only sixty-six tanks and self-propelled assault guns, twelve per cent of the tank army's original strength. The army's personnel strength was extremely low, at thirty to thirty-five per cent of officers in corps staffs. The brigades and corps had lost seventy-five per cent of their communications means.[40] With permission from the Front commander, General P. A. Rotmistrov, commander of the 5th Guards Tank Army, created in each corps composite brigades consisting of two or three tank battalions and one motorized rifle battalion. These were combined into an army detachment.

The command and control of the composite army detachment was carried out by the staff of the 5th Mechanized Corps which had the greatest numerical strength. It received two high-powered radios and three medium powered ones from the 18th and 29th Tank Corps. The operations and intelligence sections were reinforced by officers from the appropriate sections of the army staff. The staffs of the composite brigades were created by unifying several staffs which had lost a significant portion of their men and equipment. For example, the commander of the 25th Guards Tank Brigade headed the composite brigade of the 29th Tank Corps. His staff had been formed from the staff officers of two other brigades. The operations and intelligence sections had been transferred from the staff of the 29th Tank

Corps to the staff of the composite detachment. Engineer support was directed by the chief of the engineer troops from the corps.

These measures allowed Rotmistrov to successfully continue the offensive with a strong unit forward and two phantom corps for deception, while simultaneously stopping parts of the army for restoration. This brilliant risk and improvisation belies the conventional notion that the Red Army advances came as a result of a massive superiority in personnel and equipment. Field Marshal von Manstein, commander of the German Army Group South, in his memoirs leaves the reader with the impression "the main danger in this sector lay in the superior strength of the pursuer - particularly of his armor which was able to thrust into the midst of our forces while they were actually on the move."[41] Typical of German accounts, there is little credit for true Red Army military prowess - manifesting itself in flexibility and agility.

In other instances, composite detachments were organized to carry out unforeseen tasks, such as repelling German counterstrike and maneuvering to a new sector in the absence of reserves and second echelon forces, for example, the 1st Tank Army in the Zhitomir-Berdichev operation. In the 2nd Tank Army in the Vistula-Oder operation and in the 6th Tank Army in the Khingan-Mukden operation, their composite detachments were created as a consequence of interruption in fuel deliveries.[42]

During the Korsun-Shevchenkovsky operation, February 1944, the 2nd Tank Army, as a consequence of the crisis situation and due to the fact that its brigades and corps were arriving by rail in the assembly areas, formed its fragmented units under a unified command. The staffs of the formations which had received army and corps units were strengthened by officers from army and corps field headquarters. The army's assault gun regiment and a number of other units were shifted to the command of a tank brigade.

Within the existing force structure for tank armies and corps, and mechanized corps, the Red Army created very agile forward detachments. The most combat effective units; often a reinforced brigade, commanded by the most experienced, innovative, aggressive officers, formed the forward detachments.[43] Assigned missions of disrupting enemy command and control, these detachments swiftly separated themselves from the main forces and seized road junctions, bridges, crossing sites and bridgeheads in river crossings, and held them until the approach of the main forces. They ensured an unimpeded advance by the main force creating favorable conditions for their offensive.

Highly mobile, they increased offensive tempo. Success resulted mainly from effective reconnaissance, surprise, and determined operations. The reconnaissance effort focused on revealing the disposition and outline of the German's forward defensive positions, weak spots in their defense, and the approach of reserves. After reconnaissance, the forward detachments with air support attacked suddenly unprepared German intermediate lines. Taking advantage of success achieved by these detachments wedged into the German defense, the army's main efforts would concentrate on narrow sectors of three to four kilometers, penetrating the German's defense with an attack from the march. With support of artillery fire and air strikes, the offensive developed in the depth reaching the flank and the rear area of the Germans' defense.

The agility of tank armies grew under more confident handling of aggressive and experienced commanders and staffs. By late 1943, tank armies began to show their potential for greater mobility and agile commitment in support of Front level operations. The regrouping of the 3rd Guards Tank Army at the Dnepr river in October 1943 is an example of the tank armies' growing battlefield agility.

At the end of September 1943, the Voronezh Front (redesignated the 1st Ukrainian Front 20 October 1943) advanced to the Dnepr River. Forcing the river in a number of places, the Front secured two major bridgeheads in the vicinity of Lyutezh and the Velikii Bukrin. Developing the offensive from these bridgeheads created an enveloping threat from the north and south to the German forces centered on Kiev.

Twice in October renewed offensives from the Bukrin bridgehead and adjacent areas by the 3rd Guards Tank, 40th, and 27th Armies were unsuccessful. The Stavka and Voronezh Front commander began to look for alternatives to force the Dnepr River.

With the addition of the 5th Guards Tank Corps to the Lyutezh bridgehead in early October, the northern bridgehead was widened to 20 kilometers and a 15 kilometer depth. Success in building up forces in the Lyutezh bridgehead offered a possibility for redirection of the Voronezh Front's main effort. The Stavka ordered the Front to regroup part of its forces, in lieu of additional forces, for use in the Lyutezh bridgehead. Included in this regrouping was the 3rd Guards Tank Army.

At 1800 hours, 25 October, General Lieutenant Pavel S. Rybalko received a Front directive to begin withdrawal of his units on available crossings from the Bukrin bridgehead to the east bank of the Dnepr River.[44] Upon receipt of the directive, Rybalko assessed the mission, considered the situation, and gave preliminary instructions. The Front's concept of operation transferred the occupation of the bridgehead to 40th Army by 0300 hours,

27 October. Rybalko's tank army moving to the east bank of the Dnepr had to conduct a night movement only to designated assembly areas and to recross the Dnepr into the Lyutezh bridgehead. For fulfilling the assigned missions was a three stage plan: withdrawal from the Bukrin bridgehead, direct march, and occupying final jump off positions for the attack.

The withdrawal from the bridgehead required the use of deception in the creation of dummy fighting positions along the front lines. Using partially destroyed and unrepairable combat vehicles and false gun tubes, the forward positions looked fully occupied. Additionally, the corps command post radio stations were kept in place with no break in their routine transmissions. The tank army's communication signature would mask the withdrawal of staffs with their command posts.

Crossings on the Dnepr were prepared by Front engineer units. They managed four ferries of varying capacities and two pontoon bridges of thirty-ton capacity. Additionally, a thirty-ton capacity wood pile bridge located at Zarubentsy offered another combat vehicle passage.

Corps and bridges began moving at 1900 hours 25 October. In the first half hour thirty-six tanks crossed in the area of Zarubentsy. The thirty-seventh tank, unfortunately, had run off the bridge and fallen into the water. The accident knocked out five pile supports. Through the night, units passed to the east bank only on the pontoon bridges and the ferries. Some sixty-one tanks and wheeled vehicles of the 9th Mechanized Corps, 91st Separate Tank Brigade and 50th Motorcycle regiment made the crossing the first night.

The following night, eighty-two tanks and all the motor transports of the army crossed. Fortunately, a thick fog lingered during the morning and tanks continued to move until 1100 hours. At 0800 hours 27 October, the bridge repairs were completed, but after crossing two vehicles the bridge was out of action again. By 0600 hours, 28 October the tank army was completely out of the bridgehead and moving north. Some units were strung out, but all the corps headed north.

The slow crossing of the Dnepr resulted from a lack of time for required preparations and little experience by personnel in river crossing operations. The commandant service which provided the road guides to the crossing points was hurriedly organized by corps and brigade commanders in a decentralized manner. Untrained personnel allowed combat equipment and vehicles to stack up at one crossing while making other crossings lose time by waiting for units.

From the beginning the main force march received focused attention for timely arrival in selected areas. Commanders, staffs and political officers and corps commanders could be found directly in the units rendering personal help to the commanders of brigades and battalions at the crossings and difficult sectors of the march routes. As a result of this concentrated command effort, a large part of the force and equipment managed the march in a short time.

Crossing the Desna River in the second stage of the tank army's move was organized better than the first crossing of the Dnepr. Beforehand, the Front had prepared wooden bridges in the area of Letki without delaying the army's advance, while heavy self-propelled artillery vehicles crossed on a ford south of the bridge. The army recrossed the Dnepr into the Lyutezh bridgehead on ferries and bridges. The first night 110 tanks crossed, on the second -185, and on the third the rest of the tanks.

At night on 4 November, 3rd Guards Tank Army was committed to the fight. Developing the offensive, the tank army in a few days advanced toward the western edge of Kiev. The crossing of the Dnepr was the first big test in river crossing for the Red Army as it began its drive out of Russia to the west. Agility lessons in movement through the environment and coordination at the operational level were learned by the tank army. From late 1943, Red Army offensives would continuously confront river crossings as their advance came to many of the well-known rivers of Europe.

In November 1943, five tank army commanders and their political officers were called to Stalin's office to discuss tank army formations. Drawing on the commanders' experience, Stalin wanted to work out any necessary changes to a formation that would spearhead his main efforts in upcoming large offensive operations.

General Pavel S. Rybalko, the extremely aggressive, hard-hitting commander of the 3rd Guards Tank Army, suggested withdrawing the hospitals, medical detachments, and road-bridge battalions from the tank army. These elements, he believed, "weighted down" his units and caused unnecessary troubles. Their support could be received from the rifle army and Front sectors in which the tank army was committed.[45]

General Mikhail E. Katukov, the methodical and calculating commander of the 1st Tank Army, objected that, "Our doctors acquired experience from treating wounded tankers. Their wounds are particularly special, very often connected with serious burns."[46] Indeed, the statistical legacy for the modern Soviet military researchers implied that the greatest weight of casualties suffered by tank troops around twenty per cent was from shellshock, burns and bruisings - casualties unique to armored warfare.[47] Very clearly, the link between force structure and a unit's viability for sustained operations is understood by today's Soviet army based upon a rigorous analysis of their legacy.

Other direct connections between sustainability, and viability, could be seen in the increase in firepower, shock power, and maneuverability of the tank formations. The creation of stable command and control improved the ability to continue fighting despite considerable losses. As noted in a recent study:

"Sustainability, viability, is not just a question of logistics. It is the active prosecution of the battle, the provision of the proper equipment and the maintenance of supplies and reserves, the careful structuring of forces, the skillful choice of operation and the skill with which it can be conducted that in his influential view most contribute to an army's ability to sustain itself in battle. It is above all a positive approach: the more effectively the enemy is hit, the less casualties he will inflict and therefore the longer our operations can be sustained."[48]

By the end of 1943, the Red Army had a tank army structure that would endure to war's end. With experience and increased mobility of tank armies, the Red Army's battlefield agility shifted from a tactical to operational level. In 1944, massing of armored formations created powerful Front mobile groups and contributed significantly to Red Army victories. Placement and focus of tank armies consistently produced desired correlation of forces at the operational level. The Lvov-Sandomierz operation illustrates what tank armies in conjunction with other mobile groups could achieve.

In the summer of 1944, the Red Army struck successfully against Belorussia and the Northern Ukraine. In the wake of the Belorussian Operation, in July 1944, the 1st Ukrainian Front hurled itself against the German Army Group Northern Ukraine employing two simultaneous main efforts - a first for the war. The assault groupings in each sector had two rifle armies to penetrate the German defense and armored forces to exploit success. On the Rava-Russkaya sector a tank army and cavalry-mechanized group (CMG) followed, and on the Lvov, two tank armies and a CMG.

The Lvov-Sandomierz operation involved a large scale regrouping of forces as an operational level deception. Since in the previous combat operation the main Front forces were on the southern wing, half of these forces had to be regrouped over distances of 100-300 kilometers and more, including the 38th Rifle Army, all three tank armies, and four artillery divisions. Much of this repositioning was masked by deception activities. Although the deception was not entirely successful, its effect allowed the Front to achieve some dramatic surprises.

A unique feature of the Lvov-Sandomierz operation was the successive commitment of two tank armies to battle in the zone of advance of a single rifle army. At the moment of planned commitment of the tank armies, the 38th Army's zone of advance in which the 4th Tank Army was to be committed had been severely blunted by a two panzer division counterattack, and the 60th Army, where the 3rd Guards Tank Army was to be committed, together with its forward brigades broke through the defenses on a 6 kilometer wide and 20 kilometer deep corridor. The counterattack into the flank of the penetration threatened to jeopardize development of the offensive in the sector of the main thrust. In order to maintain the breakthrough tempo, the Front commander committed initially the 3rd Guards Tank Army through the narrow corridor and then followed with the 4th Tank Army.

Supporting the commitment of two tank armies, the 60th Army was reinforced with artillery and ordered to hold firmly the corridor flanks. Significant air power was assigned to neutralize German units in front of the tank armies and on their flanks, especially in the location of counterattacking panzer divisions.

In the Lvov-Sandomierz operation the Red Army gained great experience in moving forces to encircle and eliminate large German groupings. Characteristically, at the start of combat operations, the Front did not hold an enveloping position as was the case at Stalingrad and in the Korsun-Shevchenkovsky operations. The conditions for encirclement were created in the course of the offensive after breaking through the tactical defensive zone of two distant sectors. The prompt commitment of mobile forces to the penetration in each sector, and their rapid maneuver from two sides toward the flanks and the rear of the German Brody grouping ended with the encirclement and destruction of eight German divisions.

The operation featured execution of aggressive, dynamic maneuvering, particularly by armored and mechanized forces. Expansive maneuvering of mobile forces in the course of the operation made it possible to quickly concentrate necessary forces in selected sectors and make surprise attacks against the Germans. For example, having recognized the threat of encirclement for the Brody grouping, the command of the Army Group Northern Ukraine endeavored to pull it back. However, a rapid maneuver by the CMG No. 1 (25th Tank and 1st Guards Cavalry Corps) into the rear of this grouping prevented the German units from avoiding encirclement. Later, after an unsuccessful attempt to capture Lvov by a hasty frontal attack without a halt, the 3rd Guards Tank Army swept north around Lvov, emerged in the area further west of Lvov, and then with a portion of its forces attacked the Lvov grouping from the west. At the same time, the 3rd Guards Tank Army left one corps to help with the encirclement of German forces at Brody and used the third corps to continue advancing west into the operational depth. General Rybalko had his tank army moving in three different directions at the same time!

Characteristic of the operation was the repeated shifting of efforts from one sector to another. For example, when it finally became clear that the breakthrough could be carried out successfully in 60th Army's area, the Front commander committed the 4th Tank Army to battle behind the 3rd Guards Tank Army. This prompt decision to shift efforts prevented the enemy from substantially altering his plan of operations and avoided defeat in the Lvov sector.

In the course of the Lvov-Sandomierz operation 1st Ukrainian Front forces crossed a large number of major rivers including the Western Bug, Dnestr, San, Vistula, and others. At times assault groupings reached two rivers simultaneously. Successful crossing by forward units was achieved by surprise, by the crossing without a halt and by careful preparation and skillful allocation of crossing and engineer equipment to mobile forces. Tactical commanders' initiative also contributed to initial crossings success. Traversing water obstacles was accomplished with more ease than in earlier operations. In the successful Vistula river crossing, massed use of air defenses and prompt relocating of aviation helped ensure success.

General P. S. Rybalko's 3rd Guards Tank Army in 1st Ukrainian Front operations in Silesia in January 1945 are an excellent example of redirecting a tank army on the move. The operation exhibited the Red Army's tank force mature capability to accomplish agile maneuver. The tank army successfully advanced on Breslau while the Front's left flank began to seriously lag in front of stubborn German resistance which organized a strong defense in the Siliesian industrial region, which had tremendous economic importance to both sides. In order to disrupt the enemy plan to withdraw their forces behind the Oder river and speed up the movement of the 5th Guards Army to the river, the Front commander, General I. S. Konev, decided to turn 3rd Guards Tank Army to the south and have it strike the rear of the German concentration.

At 0300 hours, 21 January, General Konev ordered the 3rd Guards Tank Army to strike southwesterly toward Namslau and Oppeln, instead of attacking the railroads and highways leading to Oppeln from the east and southeast, and to prepare to force the river behind the 5th Guards Army to attack further to the west.

Midday on 22 January, when the 3rd Guards Tank Army was already attacking westward, Konev refined the army's mission and ordered it to force the Oder from the march, without waiting the approach of the 5th Guards Army, and seize bridgeheads near Oppeln. The 3rd Guards Tank Army changed direction of its offensive by 90 degrees and advanced by nightfall on the second day a depth of 70 kilometers. Owing to the rich experience acquired by Rybalko, his staff and subordinate commanders, the army accomplished the turn to the new axis in a very short time. Already, by 0500 hours, 21 January, units received their missions, and at 100 hours, they went on the offensive.

Rybalko decided to attack toward Oppeln with two tank corps in first echelon and his mechanized corps in second echelon. The reinforced 55th Guards Tank Brigade, in the area of Landsberg, was designated as forward detachment for the 7th Guards Tank Corps, and the reinforced 51st Guards Tank Brigade, which had previously been in the corps second echelon, was assigned as forward detachment for 6th Guard Tank Corps. The army dispatched reconnaissance groups to lead the advance.

By evening 22 January the tank corps had reached the Malapane river. Operating in a heavily forested mountain area, overcoming mined barriers and fierce German resistance, they forced the Malapane at night and drove swiftly toward the Oder. On the morning of 23 January, the motorized elements of 6th Guards Tank Corps seized Eyzenau and Groshovets. On 23 January, 7th Guards Tank Corps reached the Oder south of Oppeln and cleared German troops from the eastern bank.

The 3rd Guards Tank Army's turn to a new axis required reconnaissance be organized anew. Seeing the battlefield, as war experience taught, became a fundamental aspect of agility. For sustainment, troop units and rear services also regrouped. In order to fulfill the Front commander's order, the tank army had to move 70-80 kilometers and its forward detachments approximately 100 kilometers by midday on 24 January.

At 0310 hours 24 January, army troops were assigned new missions, and by dawn shifted their offensive for the second time to a new axis. Rybalko decided to place the 9th Mechanized Corps in first echelon and move the 6th Guards Tank Corps to second echelon. The first echelon corps led with forward detachments.[49] The 3rd Guards Tank Army clearly demonstrated, a year and half later, an agility comparable to that of the 10th Tank Corps at Kursk.

The Red Army built sustained operational viability for tank armies by increasing their mobile combat power, that is, tank destroyer units, self-propelled artillery, air defense, multiple rocket launcher and engineer units during the last years of the war.[50]

Command and control for armored forces to enhance agility manifested itself in command post configurations, staff procedures, and efficient parallel planning. For centralized, continuous control, Red Army unit commanders and staffs employed operational groups, main and rear command posts. Rapid displacement of command posts was indicative of well-trained, disciplined command groups. The ability to routinely operate at night implied individual and team expertise. "From the experience of twenty-five offensive operations by tank armies, with an average duration of continuous combat operations of twelve-thirteen days, the command post changed its position ten-twelve times, that is, almost daily, and the operational groups two-three times daily."[51] Depths of advance of 60-80 kilometers and more forced command posts to move at least twice a day and this required spending up to 48-50 per cent of the time of the day.[52] In this instance command and control functioned on the move, and radio capabilities did not fully satisfy needs. Consequently, the stability and continuity of command and control often dropped sharply.

A remedy for this problem was found by creating two positions for communications equipment in tank armies from the summer of 1944 onwards. This was done in order to reduce the time losses in setting-up and taking-down command posts from the total time of their move. Time was also reduced by ferrying staff officers and other key personnel on small aircraft to the new area and previously set-up command posts. This technique of flying the staff was used by the 3rd Guards Tank Army on the night of 20-21 July 1944, in the Lvov-Sandomierz operation.

Frequently, command posts changed locations by leapfrogging. The main command post shifted to the location of the operational group and the rear CP to the main CP. The command posts of armored formations shifted in long jumps (15-30 kilometers) and command posts were changed often at all levels. Existing communications restricted the number of command post moves to not more than two or three a day. Since corps operational groups were more mobile, their command post moves involved less time. For example, the operational group of the 9th Tank Corps in the Vistula-Oder operation halted for twenty-thirty minutes every hour and a half of movement.[53]

Map graphics and operational concepts supported agility as well. Soviet map graphics were (and are) designed for time and space control of offensive tempo. They show a battlefield with a clarity notably greater than NATO symbology. They better illustrate location, disposition, activity, composition, density and successive actions. Missions and objectives are depicted in directions and lines. To allow freedom of action and provide the latitude necessary for agility, armored forces did not receive sectors in the operational depth, but rather indicated direction of operation.[54]

By reviewing Red Army tank operations during the last two years of the Second World War, one is struck by the agility shown in these operations. Repeated regroupings of forces from one sector to another, broad encircling maneuvers, adjustments or changes in combat missions due to the rapid, frequent, and sometimes drastically changing situation, were all present.

This Red Army's legacy in battlefield agility creates a font of practical wisdom which the Soviets may draw upon when making decisions and taking actions today. It identifies a perceptible relationship among important factors contributing to agility. For a force to be agile it must be able to move through the environment with tailored support from artillery, air defense, engineers, intelligence, and aviation, as well as logistics and maintenance. Agile operations require concentric command that continually focuses the combat power at the right place and time to exploit enemy weaknesses and vulnerabilities, or creates weaknesses and vulnerabilities by its actions. Wide latitude for agility requires centralized command that selects the types of engagements and allocates the resources, and it entails loosely tethered control that coordinates and projects forces without formed sectors into the operational depth. Sustaining force viability operationally and logistically must be understood. The implications of force structure, maintenance and recovery, and focused objectives are essential. Finally, for a force to be agile, it must see the battlefield in terms of both terrain impediments and enemy force, identifying weaknesses, reserves, and direction of fighting.

Unlike the west with only a vague inkling of General George Patton's armored warfare legacy, the Soviet Army has rigorously researched the experiences of its armored warfare commanders, Katukov, Rotmistrov, Rybalko, and other tank commanders. That analysis of tactical and operational indices chart what is empirical on the uncertain battlefield. Their expectations based on understandable and credible experience shape the structure and composition of their forces today. Concerning the altering implications of practical experience, Marshal of the Soviet Union I. S. Konev wrote,

> "When I recall the war and compare its different stages, I feel that at times we underestimate the path we traversed in mastering the art of war during those years. In the fourth year of the war we thought it natural that we could perform such

combat missions which, if mentally transferred to the first
stage of the war, would be considered incredibly difficult and bordering on the impossible."[55]

Not content to let a concept of 'fog of war' or 'friction' underwrite the unknown, their utilitarian approach to past experience explores the limits of uncertainty for practical application of today's problems. Current Soviet armored theorists recognize the contrast of past with present-day conditions, and that the nature of armored operations will be influenced by fundamentally new factors and technology. Soviet armored forces possess a rich legacy for battlefield agility and seem ready for the spacious, fluid, non-linear conditions of warfare in the twenty-first century.

NOTES

1. *Field Manual 100-5, "Operations"* (Department of the United States Army, Washington, DC), May 1986, p. 16.
2. Ibid.
3. Ibid.
4. A. I. Radzievskii, *Tankovyi udar* (Moscow: Voenizdat, 1977), p. 7.
5. A. Eremenko, *The Arduous Beginning* (Moscow: Progress, 1966), pp. 9-10.
6. P. A. Rotmistrov, *Vremya i tanki* (Moscow: Voenizdat, 1972), p. 63.
7. Eremenko, pp. 15-17.
8. Rotmistrov, p. 71.
9. M. Dorofeev, "O nekotorykh prichinakh neudachnykh deistvii mekhanizirovannykh korpusov v nachal'nom periode Velikoi Otechestvennoi Voiny", *Voenno istoricheskii zhurnal* (hereafter *Vizh*), 3(1964), p. 34.
10. Ibid., pp. 35-36.
11. Ibid., p. 37.
12. V. A. Anfilov, *Nachalo Velikoi Otechestvennoi Voiny* (22 Iyunya - seredina Iyulya 1941 goda) (Moscow: Voenizdat, 1962), p. 74.
13. Eremenko, p. 59.
14. Anfilov, p. 154.
15. *The Halder Diaries*, Vol 2 (Boulder, CO: Westview Press, 1976), p. 173.
16. Ibid., pp. 175-6.
17. C. N. Donnelly, *The Sustainability of the Soviet Army in Battle* (Sandhurst, Camberley: Soviet Studies Research Center, 1986), p. 129.
18. Heinz Guderian, *Panzer Leader* (London: Futura, 1974), p. 234.
19. Ibid., p. 235.
20. P. A. Rotmistrov, "Boevoe primenenie bronetankovykh i mekhanizirovannykh

voisk", *Vizh,* 12 (1967), p. 39.

21. Ibid., p. 40.

22. *Sbornik materialov po izucheniyu opyta voiny, No. 6* (April-May 1943), General'nyi Shtab Krasnoi Armii (Moscow: Voenizdat, 1943), p. 57.

23. Ibid.

24. Ibid., p. 61.

25. *Sbornik materialov po izucheniyu opyta voiny, No. 8* (Aug-Oct 1943), General'nyi Shtab Krasnoi Armii (Moscow: Voenizdat, 1943), p. 50.

26. Ibid., p. 54.

27. Rotmistrov, *Vizh* 12 (1967), p. 40.

28. Ibid., "Prikaz NKO No. 325 ot Oktyabrya 1942 g," *Vizh* 10(1974), p. 68.

29. Radzievskii, p. 25.

30. I. M. Kravchenko, V.V. Burkov, *Desyatyi tankovyi Dneprovskii* (Moscow: Voenizdat, 1986), p. 72.

31. Ibid., p. 79.

32. Ibid., p. 80.

33. Ibid., p. 83.

34. Ibid., p. 90.

35. Ibid., p. 92.

36. Ibid., p. 96.

37. Ibid.

38. Ibid., p. 102.

39. *Sbornik materialov No. 8,* p. 67.

40. A. I. Radzievskii, "Poddershanie i vosstanovlenie boesposobnosti tankovykh armii v nastupatel'nykh operatsiyakh," *Vizh* 3 (1976), p. 19.

41. Erich von Manstein, *Lost Victories* (CA: Presidio, 1982), p. 469.

42. P. P. Tovstukha and R. M. Portugal'skii, *Upravlenie Voiskami v Nastuplenii* (Moscow: Voenizdat, 1981), p. 165.

43. F. D. Sverdlov, *Peredovye otryady v boyu* (Moscow: Voenizdat, 1986), pp. 21-2.

44. S. Alferov, "Peregruppirovka 3-i gvardeiskoi tankovoi armii v bitve za Dnepr (Oktyubr' 1943 g.)," *Vizh,* 3 (1980), p. 16.

45. M. E. Katukov, *Na ostrie glavnogo udara* (Moscow: Voenizdat, 1976), p. 254.

46. Ibid., p. 255.

47. Donnelly, p. 64.

48. Ibid., p. 45.

49. *3-ya gvardeiskaya tankovaya* (Moscow: Voenizdat, 1982), p. 215.

50. "Otvety na pis'ma chitatelei," *Vizh,* 9 (1973), p. 123.

51. Tovstukha, p. 176.

52. Ibid., pp. 176-7.

53. Ibid., p. 178.

54. Radzievskii, *Tankovyi udar*, p. 46.

55. I. S. Konev, *Sorok Pyatyi* (Moscow: Voenizdat, 1966), p. 111.

CHAPTER 7

Red Army Indicators

Knowing the enemy is not a modern phenomenon in warfare. Evidence of its existence can be traced to man's first endeavors at organized tribal warfare. The great battle captains of history strove to outguess and outmaneuver their adversaries by determining as accurately as possible a picture of their capabilities, strengths, and intentions. For most of recorded history, information was gathered solely by agents and military observers, providing specific facts to determine enemy intentions.

Today, information is gained through a number of sources based on technical systems, almost to the exclusion of agents and military observation. To avoid neglecting historical precedents, modern intelligence collection methods must ensure that the role of human sources is not neglected for more technologically-oriented collection sources, such as radio intercept or overhead photograph radars. In building estimates on enemy intentions, armies use their collection capabilities to find indicators of enemy intentions. One learns, for example, to identify 'classic' indicators of attack: artillery positions well forward, massing of tanks and troops and logistical build-up. While these activities lend themselves to easy collection by technical systems, one lacks the certainty that they are not being manipulated for deception, and these activities cannot specifically identify when the attack will begin. If these 'classic,' or macro, indicators are not enough for understanding the enemy, modern intelligence systems must then look to a source that has continued through the ages and still holds contemporary significance.

In *The Art of Warfare*, Sun Tzu, a Chinese general around 500 B.C., uttered the oft repeated caution ". . . one who knows the enemy and knows himself will not be endangered in a hundred engagements. One who does not know the enemy but knows himself will sometimes be victorious, sometimes meet with defeat. One who knows neither the enemy nor himself will invariably be defeated in every engagement."[1] The famous general also identifies 'signs of significance' communicating enemy capabilities and intentions as illustrated by the following examples:

> If the enemy's troops march up angrily and remain facing yours for a long time without either joining battle or taking themselves off again, the situation is one that demands great vigilance and circumspection.
>
> When an army feeds its horses with grain and kills its cattle for food, and when the men do not hang their cooking pots over the campfires, showing that they will not return to their tents, you may know that they are determined to fight to the death.
>
> When there is much running about it means that the critical moment has come.
>
> If their flags and pennants move about, they are in chaos.
>
> If the officers are angry, they are exhausted.
>
> Those who stand about leaning on their weapons are hungry. If those who draw water drink first, they are thirsty. When they see potential gain but do not know whether to advance, they are tired.[2]

Sun Tzu's ageless perspective is at the core of human endeavors in warfare. All armies must time their marching and fighting around needs of soldiers and machines of war. The soldier will always exhibit certain human frailties; i.e., hunger, fear, and demoralization. Fighting machines will always exhibit limitations which need refueling, rearming, reequipping. Human sources for the collection of such micro-indicators is critical in discerning specific enemy intentions and capabilities.

Since Sun Tzu, immense progress in technology has contributed significantly to intelligence collection. During World War I, the application of technology to intercept wireless communications and use of aerial photography was rudimentary in development. Despite the new collection devices, observation and patrolling by frontline troops proved necessary to identify enemy machinegun cupolas, dugouts, belts of wire and enemy trench lines.

With World War II, armies increased the sophistication of aerial reconnaissance and use of overhead photography. More signal units were dedicated to intercept radio traffic and famous code breaking of ciphered transmissions. While agents could still deliver valuable information, the pace of tactical operations limited their use. Reconnaissance and patrols were major sources of information.

In the latter part of World War II, as cited from Cornelius Ryan's book,

The Last Battle, the German General of the Army Group Vistula, Colonel General Gotthard Heinrici, was highlighted for his ability to ascertain the exact time of the Red Army attacks. Since Heinrici never published his memoirs, he remains a relatively unknown German commander from the Eastern Front. He possessed, however, an ability to pinpoint the attack hour, enabling him to evacuate frontline positions prior to massive Red Army artillery preparations and conduct his defense with greater flexibility.

Almost daily, General Heinrici flew the enemy frontline trace to observe troop activities and artillery deployments. At night, he studied prisoner of war interrogations and other intelligence reports for clues which would help ascertain time of attack. Heinrici was not fooled by the macro indicators routinely manipulated by Russian commanders at this stage of the war. At one point along the front, the Red Army would lay down a short artillery barrage, and in another sector, they would launch limited attacks. For years, the Red Army had used these same feints and ruses as preludes to main attacks. Heinrici's concern was when to order his troops out of front line positions back to the second line of defense.

What were the essential elements of information which gave Heinrici a prophetic, almost clairvoyant, estimate of Red Army intentions and time of attack?

In April 1941, reassigned as Commanding General of the XLIII Army Corps which conducted operations on the Loire and saw Channel defense in Normandy, General Heinrici with his corps moved to Poland. In June 1941, he participated in the German invasion of Russia, fighting battles for Byalistok, Gomel, Kiev, Bryansk, and Moscow.

General Heinrici's tactical mastery against Red Army attacks begins to emerge with his handling of the XLIII Army Corps in December 1941, in defensive battles around Tula during the Red Army Moscow counteroffensive. The XLIII Corps, as part of the Fourth Army, was exposed to Red Army forces exploiting a wide gap west of Tula on 20 December. The Red Army units moved around XLIII Corps open flank to attack Kaluga, Fourth Army's railhead and supply base. Continuing the assault in a driving snowstorm, the Russian 49th Rifle Army broke through the German Fourth Army, splitting off and encircling the XLIII Corps. Heinrici's Corps on the southern flank of the Fourth Army possessed too few troops to cover the flank. The desperate situation continued for a couple of weeks. Finally, on the night of 5 January 1942, General Heinrici, who liked to know the big picture, called Fourth Army Chief of Staff and asked whether the army was being deliberately sacrificed and being treated like Red Army units during the summer encirclements. Heinrici's abusive charge elicited an immediate response from Field Marshal Guenther von Kluge, the Army Group commander, who had been reading accounts of Napoleon's 1812 invasion of Russia and was fearful, as many other German officers, that a sudden morale collapse would turn into a rout as it did for the French Army retreat nearly a century and a half before. Field Marshal von Kluge transmitted to the Fourth Army that the leadership should not fold and that he would not forsake the Army.

German units had been overextended in logistics, landmass covered, and limits of human endurance for their soldiers. Their strongpoint defense during the winter of 1941-42 was a tactic of weakness, dictated by German unit weakness, the harsh Russian winter, the worst in 100 years, and Hitler's standfast order.

By mid-January 1942, the Fourth Army had been allowed to retreat from threatened flank and rear. General Heinrici, the senior officer at the front since the army headquarters was located sixty miles to the rear, claimed the troops' confidence in their leadership was collapsing. Orders to hold at all cost had been read to them only a day before, and now they were being told to pick up and move. This was unstable, unreliable leadership to Heinrici's style of fighting. His soldiers must have absolute confidence in his commands when to hold positions at all cost and when to withdraw under pressure. Otherwise the commander could not shape the battle and control the fight.

Commanding units since the beginning of the war, Heinrici possessed a low key style of command, without the dash or flare of a Rommel or Guderain, also contributing to his postwar anonymity. Short, slightly built, with blue eyes, fair hair and trim moustache, Heinrici looked more the part of a college professor as he looked over the top of his reading glasses. He did not swear, he did not drink, he did not have a mistress, was colorless and not much material for anecdotes. His threadbare uniforms added to the deceptive image that hid a relentless battlefield commander. Subordinates, however, remember his unique quality of forcing his commands and decisions on them. He strove to stimulate and stretch not only his subordinates, but also superiors. Sometimes argumentative, he had an aura of old Prussian arrogance which made close relationships impossible and accented an aloofness in manner. Hard and gnarled, the short Prussian officer of the old school received from his subordinates the nickname, "*Unser Giftzwerg*," meaning our poisonous dwarf. He was reflective with an eye for precision. The military historian, Earl Ziemke, called Heinrici the "prophet of catastrophe," but he recognized that Heinrici "had never failed to bring his corps out of the tightest spot."[3]

The Fourth Army commander, General Ludwig Kuebler, found the strain of fighting on the Eastern Front too great. On January 26, 1942, General Heinrici received command of the Fourth Army in bitterly cold temperatures at minus 42 degrees Fahrenheit. Climatic elements were an enemy to both armies. However, the opportunity was just the situation in which Heinrici thrived. Always looking for a better idea and room for improvement, he displayed a natural propensity for organization.

The situation for Fourth Army turned around under General Heinrici. Upon taking the Army, he withdrew a small detachment from an awkward position it was holding. He quickly received a warning conveyed through Field Marshal von Kluge that a similar move would result in a court martial for him. Such rebukement would feed frustration and resentment in Heinrici; he wanted a free hand to exercise his defense, to experiment and improve the defense. He worked to close gaps and conducted counterattacks with artillery to stymie Russian advances through knee-deep snow. He would rebel against being too tightly controlled.

At the end of January, the Fourth Army's supply route was cut for four days. Desperately counterattacking, Heinrici's army barely covered the sector, but by 3 February, his forces bridged the gap between Fourth and Fourth Panzer Armies. He was gaining confidence in handling his forces and understanding his enemy.

In mid-February, Heinrici wanted to shorten the army's line behind the Urga river, saving troops and strengthening his position. No one in the chain of command wanted to pass the idea up the line. On 1 March, Heinrici received an interview with Hitler, and to everyone's surprise Hitler approved the move. General Heinrici managed to maintain the new position well into spring despite being deeply enveloped on both flanks.

Photographs of Heinrici on the Eastern Front depict him in his favorite, worn sheepskin coat and short boots with World War I leggings which he preferred to the highly polished knee boots of his Nazi contemporaries. Clothes were for comfort, utility, and worn out of habit; one keeps clothes for years, like old friends. On 1 January 1943, Heinrici received promotion to Colonel General, primarily because of von Kluge's persistence.

In March 1943, the spring thaw and rains brought deep, black Russian mud. The black oily Russian mud, which could suck the heel off a soldier's boot, slowed to a crawl operations on the Eastern Front. Both sides settled down and prepared for the resumption of the summer offensive, resuming the annual fight for strategic initiative. In Operation Citadel July 1943,

Heinrici's army played a backseat to the massive panzer forces which conducted offensives on the northern and southern faces of the Kursk salient.

Late summer-fall 1943, General Heinrici commanded the Fourth Army on the sector from Rogashen to Orsha, astride the major highway from Moscow to Minsk. After evacuation of Smolensk to within 20 kilometers from Orsha, his army met a series of strong Red Army offensives, beginning in October and continuing until December.

Heinrici described in detail his defensive techniques and actions to Sir Basil Liddell-Hart who retold them in his book, *The German Generals Talk.* Heinrici believed the defensive problem turns on the ratio of space to force. The five successive offensives pitted Heinrici's ten divisions to hold a 200 kilometer wide sector. The Fourth Army was without reserves. Artillery was intact and a crucial factor.

The Red Army's main objective was the rail center of Orsha. Their main assault was on a frontage of 20 kilometers, astride the Moscow-Minsk highway. Heinrici estimated the first offensive employed twenty to twenty-two divisions. "To meet this assault, I used three and a half division to hold 20 kilometers, leaving six and half to hold the remainder of the wide sector."[4] The Russian offensives each lasted five or six days. The Red Army units attacked routinely three times a day. The first assault came around 9AM after a huge artillery preparation. The second came between 10-11AM; the third, 2-3PM. "It was almost like clockwork," Heinrici recalled.[5] This characterization of Red Army offensives plays to a stereotype that is inconsistent with the results of operations on the Eastern Front. Heinrici's army was in a secondary sector and his forces never dealt with major Red Army breakthrough operations reinforced by large tank armies and armored corps.

Heinrici attributed his success to three factors: narrow division sectors; powerful artillery groupings; and replacement on average of a fresh battalion per division per day to replace casualties. "Of course, a necessary condition of success was to discover the actual intended day of the assault . . ." he related.[6]

In the winter of 1943-44, in the vicinity of Orsha and Smolensk, Heinrici held against terrific odds. The Russians had thirty-six divisions to his three and a half. His army fought off five all-out offensives, held losses on a 12 kilometers front to less than two kilometers. He received the decoration of oak leaves to the Knight's Cross.

Although his tenacity in defense was winning a quiet reputation, Heinrici never gained full support of the Nazi leadership. With pressure from that leadership, Heinrici, who suffered from various stomach disorders which he attributed to poison gas in World War I, was placed on the nonactive list by Hitler because of "ill-health." General Kurt von Tippelskirch replaced Heinrici as commander of the Fourth Army. He appreciated the value of an elastic defense, but also knew the disastrous consequences of being unable to practice it to an adequate extent.[7] Three weeks after assuming command, the front folded under a major Red Army offensive in the summer of 1944.

During his convalescence in Karlsbad, the Russians broke through the German Eastern Front. Heinrici went to see Guderian. The Army lost a number of generals in the destruction of the Army Group Center by the Red Army in June and July 1944, so Heinrici was called back to duty in August. He assumed command of the First Panzer, and he was also in command of the First Hungarian Army conducting operations in Central Carpathians and Upper Silesia. The combined force was called Heeres Gruppe Heinrici.

The First Panzer Army defended Dukla Pass through the Carpathian mountains. The Red Army finally made progress in the pass by the beginning of October 1944 - partly because Hitler had taken a panzer division for attacks elsewhere.

The Hungarians were trouble. Despite personal efforts at rapport, Heinrici recalled giving the Hungarian General three bottles of cognac. The following day the Hungarian General and his staff defected to the Russians. Heinrici's group retreated through Silesia in early 1945.

In this later stage of the war, Heinrici further developed his defensive methods. He withdrew troops from the first line of defense under the cover of night to the second line usually two kilometers behind. This avoided the awesome and powerful destruction of the Red Army artillery preparations with little damage and casualties. After the attacking ground forces hit air at the first line and lost momentum in the assault, they encountered coordinated, intact resistance. The best time for counterattack was the period of confusion when the attack had not yet consolidated its position or reorganized its forces. Timing was critical. Heinrici found self-propelled guns one of the greatest value in applying his defensive tactics.[8] He proudly claimed to have won all his defensive battles and did not have to call on higher commands for reserves. He told Liddell-Hart that he believed in the superiority of the defense, but believed Hart missed the mark in claiming an attacking force required a three to one advantage. Heinrici, from his experience, believed the attacker needs at least six or seven to one numerical advantage to succeed against a well-laid defense. In his defensive battles in late 1941, Heinrici's units held at odds of eighteen to one.[9]

On 21 March 1945, Generaloberst Heinrici was placed by Hitler in command of Army Group Vistula on the Oder river front from Fuerstenberg to the Baltic Sea. Hitler did not want to give the command to Heinrici but relented at Guderian's insistence. At the age of fifty-eight years, Heinrici, considered by many to be a master of defense, would bring to bear over forty years of military experience. Heinrici had the awesome responsibility of defending Berlin from the rapidly approaching Red Army.

His experience on the Eastern Front allowed him to keenly judge information. He needed intelligence of a greater detail than the placement of artillery well-forward and assembly of large tank formations or troop concentrations. He needed confirming information which was not easily manipulated for deception.

Although Cornelius Ryan does not give any specific clarification of Heinrici's secret, David Kahn in *Hitler's Spies* offered insights into specific factors that German military intelligence called "mosaic work." Mosaic work, done by the Ic - G2/S2 equivalent - consisted of collating separate and varied intelligence reports into a coherent estimate. German intelligence placed great value on verbal evidence from prisoners of war and radio intercept reports. "But the greatest volume and the most certain information came from the physical observation of the frontline troops," Kahn concluded.[10]

The German General Staff Section, Foreign Armies East, developed a pamphlet, *Troop Ic Service in the East,* which listed observations which would indicate Red Army intentions. Some of these macro-indicators included probing German lines by reconnaissance and raiding patrols, increased partisan and spy activities behind the lines, and increased aerial reconnaissance activity.

The place of probable attack was betrayed by increased radio nets, greater depth in troop staging areas, and massing of artillery, tanks and troops. A defensive intention was keyed by the disappearance of known units, feints and a decrease in artillery fire and air activity. Again, many of these macro-indicators lend themselves to use for deception. Macro-indicators could not have been the clues which guided Heinrici to his unerring time predictions.

The Foreign Armies East pamphlet notes that the Red Army displayed human tendencies, micro-indicators, which were as discernible as those given to us by Sun Tzu. Some of these 'signs of significant' for German military intelligence and particularly General Heinrici were:

> The narrowing of unit frontages for attack could be determined by observation of reliefs and changing sentry routes.
>
> If Russian intentions were to attack, the soldiers wore helmets; if

defensive, they wore caps.

Intentions of attack noted by nervous behavior of soldiers with movement across areas under fire, indicating new arrivals.

Prior to any significant marshalling of troops there were changes in mess and guard hours.

New racial faces and languages in frontline troops indicated new arrivals.

The issuance of liquor prior to the attack.[11]

This information was collected only by the frontline troops in contact, even if only through the sights on their rifle, with the enemy. To determine sentry routes, guard hours and mess hours required active patrolling and continuous observation by German units in contact. The information was then correlated with prisoner of war reports and compiled into intelligence that could estimate, almost to the hour, the Red Army attack. Soviet soldier's human tendencies were apparent in other collection efforts. It is human nature to better camouflage a position that one will occupy than a position to be occupied by someone else.[12] Consequently, spotty camouflaging discipline by an exceptionally good camouflaging army alerted German aerial photo-interpreters to the thickening of a sector with new preparations for additional troops.

The Red Army's strict adherence to manuals of instruction was another revealing indicator. By meeting the exact specifications in ramparts widths and lengths, types of forward openings, fragmentation walls in preparation of fortifications, the Russians revealed the type of firing position.[13] German aerial photograph interpreters could identify the difference between a 76-millimeter artillery gun open emplacement from a 76-millimeter antiaircraft artillery position. The primary mission of direct-fire artillery was to stop and destroy attacking German tanks. The basic artillery emplacement is sunken, with slit-trench shelters for the crew, ammunition recesses on both sides of the emplacement, and a ramp for the gun in the rear. Emplacements for antiaircraft artillery are provided with circular gun platforms, an exit for the guns in the rear, two slit trenches with light overhead cover on both sides of the platform for crew shelter, and one or two recesses in the trench for ammunition storage.

Information garnered from personal observations, patrols, aerial reconnaissance, and prisoner of war reports were combined for determining enemy intentions. General Gotthard Heinrici determined when to shift his soldiers to escape artillery fire, and survive in order to inflict casualties in the subsequent Red Army attack. An information disparity on the battlefield gives one commander the confidence and knowledge to attempt a daring operation, while the other commander displays a caution which allows that maneuver to succeed. Heinrici's bold tactical maneuvers and uncanny battlefield performance were based upon his accurate reading of Red Army indicators.

NOTES

1. Sun Tzu, *Art of War*, translated by Ralph D. Sawyer, Boulder, Colorado: Westview, 1994, p. 179.
2. Sun Tzu, *The Art of War*, translated by Lionel Giles, The Military Service Publishing Company, Harrisburg, Pa, 1949, pp. 74-75; *Art of War*, translated by Sawyer, pp. 208-209.
3. Earl F. Ziemke and Magna E. Bauer, *Moscow to Stalingrad: Decision in the East*, Washington, D.C., Center of Military History United States Army, 1987, p. 163.
4. B. H. Liddell-Hart, *The German Generals Talk*, New York: Quill, 1979, p. 214.
5. Ibid.
6. Ibid., p. 215.
7. Ibid., p. 217.
8. Ibid., p. 215.
9. Ibid., pp. 215-216.
10. David Kahn, *Hitlers' Spies*, Macmillan, New York, 1978, p. 411.
11. Ibid., pp. 102, 411.
12. Ibid., p. 133.
13. Ibid.

CHAPTER 8

Repelling Counterattacks and Counterstrikes

As the German field marshal intently studied the situation map, his staff officers overheard him say to himself, "And the best of luck to you!" Field Marshal Erich von Manstein had devised a plan to deal with a bold armored thrust by Red Army General M. M. Popov's mobile group into the Russian Donbas region during World War II. In February 1943, on the eve of the third battle for Kharkov, Popov's tank attack penetrated deep into German lines, approaching to within forty miles of Zaporozhye, a major city and crossing point on the Dnepr river. This thrust threatened to break open the German Army's defense in southern Russia.

To execute his plan for eliminating this threat, Manstein first turned the 5th SS "Viking" Panzergrenadier Division west to immediately tie down Popov's armored mobile group and slow its rapid advance to the southwest. Then, the marshal directed General Hermann Balck's 11th Panzer Division and part of an infantry division to sever Popov's supply lines. Finally, General Gotthard Heinrici's XL Panzer Corps counterattacked northward to eliminate the Red Army penetration.

At a decisive juncture in the operation, German intelligence intercepted, from Popov's group, a Russian radio call for fuel: "Wheels not turning." Manstein and the German command gained confidence in their move to complete the counterstroke.

Despite stubborn resistance by Red Army tankers, Heinrici coached his subordinates, "Don't waste time fighting against villages and towns. Keep moving."[1] Popov's armored group, sliced to ribbons, soon became a collection of inert battlefield coffins.

This well-known saga from Eastern Front fighting in World War II is an example of audacious, skillful maneuver to defeat a major armored penetration. Despite many unsuccessful German counters against later Red Army armored penetrations, Manstein's destruction of the Popov mobile group has become a classic model for armored warfare in military history.

Seeking to emulate a command style and fighting resoluteness similar to Manstein's model on the battlefield, the U.S. Army's AirLand battle doctrine generates a wide range of some new, and some revised, combat actions at the tactical and operational levels of war to create a greater depth and allow commanders to seize the initiative on the battlefield. The Army's warfighting doctrine aims to throw the enemy off balance with powerful blows from unexpected and multiple directions. This style of warfare demands concentric plans and decisive actions to degrade the coherence of an enemy operation in a rapid, surprising, and violent manner. Deep operations create that depth and initiative. Deep operations target ". . . freedom of action of the opposing commanders and the coherence and tempo of his operations."[2] Continuing, the manual states, "at the operational level, deep operations include efforts to isolate current battles and to influence when, where, and against whom future battles will be fought."[3] Deep operations, successfully prosecuted, allow the seizure and retention of the initiative.[4] While wrestling initiative from the attacker, commanders and their planners at each level must consider how the threat will react to these efforts on the battlefield.

Recent Soviet military consideration for the implications of AirLand Battle are evidenced by an increase in their professional military literature on various aspects of combat operations challenged by the U.S. Army's doctrine.[5] For the Soviets, AirLand Battle deep operations represent a serious threat to initiative and tempo in their penetrations of tactical defense zones and successful movement into the operational depth. AirLand Battle attacks are either counterattacks or counterstrikes within Soviet military lexicon, depending on the level of the operation. Counterattacks conducted within the tactical zone of a defense deal with efforts to restore the tactical situation.[6] Counterstrikes are essentially operational counterattacks, larger in scale, scope, duration, and tempo, employing larger unit operations.[7]

During World War II, attacking Red Army forces discovered that the necessity to repel counterattacks and counterstrikes was a frequent phenomenon.[8] As contemporary Soviet military researchers concluded from Red Army experiences, not a single major offensive operation was conducted during which the Germans did not undertake a counterstrike.[9] In the first period of the war, from June 1941 to November 1942, besides being primarily on the defensive, the Red Army did not have the necessary men and equipment to allocate for the mission to repulse counterattacks or counterstrikes at the various command levels.[10] In addition to their lack of resources and personnel, Red Army commanders did not conduct prior planning, allocate forces or conduct the necessary coordination to counter possible alternative actions by defending German forces -particularly their armored reserves.

While Popov's demise at the hands of Manstein gives students of military history a successful example of dealing with penetrating armor thrusts, the same operation also gave Red Army leadership in early 1943, costly, but valuable, lessons. Lessons from which the Soviet High Command, in a very short time, began directing measures to deal with German counterattacks and counterstrikes. Red Army force structure evolved continuously during the war, as their leadership instituted lessons learned and modified operations. Rifle divisions, for example, increased their strength with antitank guns and artillery, while rifle armies were bolstered with support by groups of armor and artillery for combined-arms armies with greater operational flexibility and firepower. The Russian's improved infantry capability stood up well against German panzer assaults into the flanks of Red Army penetrations. Tank army formations to exploit combined-arms army breakthroughs gave Front commanders a powerful, responsive and maneuverable formation for employment against German operational reserves and counterstrikes. Concentration of firepower and management of their second echelons and reserves allowed Red Army commanders to rapidly reinforce their main body, developing promptly tactical success, precluding initiative passing to German defenders. These measures significantly contributed to sustaining deep armor penetrations, holding Soviet gains and breaking the German defense in sector.

For successfully executing their offensive operations at a high tempo and into the operational depth in the latter periods of the war, Red Army commanders found it necessary, for planning purposes, to understand German tactical and operational level defensive actions.

At the tactical level in a highly active defense, German units held each advantageous line and counterattacked violently, at the first opportunity, with multiple counterattacks. Counterstrikes, at the operational level, were aimed to check an offensive operation already underway or to disrupt commitment of Red Army mobile groups (tank and mechanized corps or tank armies) in breakthrough operations. When Red Army units broke through defenses, German commanders usually sought to hold the attacking group's advance by assailing its flanks in order to isolate the second echelon and reserves and to prevent a cohesive development and sustainment of the offensive in depth.[11]

At the tactical level, Soviet military analyses of their war experiences postulate no uniform actions by Red Army commanders for repelling counterattacks. Yet, their research does identify common traits for dealing with this tactical combat problem. Their analytic conclusions emphasized, for Soviet Army commanders, major teaching points useful for isolating planned counterattacks and counterstrikes in AirLand Battle actions.

Successfully estimating German commanders' counterattack plans gave attacking Red Army commanders a tremendous advantage. The Soviet commander and his staff required a thorough knowledge of enemy tactics and an extensive study of the enemy's disposition. Red Army commanders gained knowledge of German tactics through experience, while their exact disposition was obtained through intensive reconnaissance and intelligence collection efforts. An increasingly stereotypical response by German panzer units resulted in a predictability for Red Army commanders. German units, as a rule, launched counterattacks into the flanks and at the base of advancing forces. Red Army tactical commanders anticipated this action by allocating forces to protect and hold the flanks. A well-organized reconnaissance that was continuous, active, and timely in reporting, became a necessary ingredient for Red Army successes. Timely discovery of enemy plans allowed preparation for repelling German counterattacks. Today, the Soviet Army has made a significant investment in a variety of intelligence collection means, particularly in reconnaissance units. They seem willing to pay for combat information in blood.

Selecting the most effective method for repelling counterattacks was a critical decision by attacking commanders. Fully appreciating the dynamics of battle, Red Army commanders often selected a method in response to a particular enemy action. One method required the entire attacking force to go on the defensive in order to deliver intensive artillery and antitank fire from the halt to prevent German forces from breaking into the formation. Red Army tactical commanders identified tank approaches and covered them with antitank fire. Artillery shifted quickly, organized, and, depending on the amount of time, coordinated a hasty defense with engineer support. This course of action was the most likely choice, if the counterattacking force was numerically superior.

A second method repelled counterattacks while on the move, particularly if the counterattack was conducted by small forces in a piecemeal manner. This option usually resulted in a meeting engagement. Both sides, in a dynamic, fast-paced situation, attempted to seize the initiative. A meeting engagement resulted when reconnaissance was poor or counterattacking forces achieved surprise.

A third method for repelling counterattacks required the most flexible and agile response by penetrating Red Army forces. The directly threatened portion of attacking forces would halt to defend against counterattacking forces. While the defending portion fixed counterattackers, the remainder of advancing forces would, depending on the strength of the counterattack, attack German flanks or continue to advance. Ideally, this alternative allowed the main attack to continue, maintaining the offensive initiative.

An important technique common to all three methods sought to inflict extensive artillery strikes on counterattacking German forces. Artillery preparations of the battlefield not only attacked defensive positions in depth, but also struck counterattack groupings as a preemptive technique. In those instances when German commanders preserved their combat capability by dispersion and then counterattacked, artillery fire and ground attack aircraft strikes reached their highest pitch in the course of the operation. Artillery often played a decisive role in repelling German counterattacks. Artillery batteries, in hasty firing positions, allowed German tanks to close and fired on them at point blank range, choking-off the counterattack.

These common techniques, when worked consistently in a coordinated fashion, became an inseparable part of offensive combat and a necessity for achieving the objective by continuous offensive operations. As Red Army experience showed, successful destruction of German Army reserves and skillful repelling of their counterattacks was a decisive condition for achieving high attack tempos.[12] A distillation of lessons is succinctly written in the Soviet Army tactical primer, *Taktika*:

> "Today, attacking subunits are equipped with more powerful and longer range fire weapons and therefore, when repelling counterattacks, they seek to cause as much damage as possible to the enemy reserves as they are discovered in concentration areas and while they are moving forward. For this purpose, nuclear and first strikes are made against the enemy, with resources under the direct control of the senior commander (chief), while the attacking subunits, when conditions are right, make encounter attacks (meeting engagements) from the line of march against the reserves moving up from the rear. This makes it possible to repel enemy counterattacks without substantially reducing the momentum of advance."[13]

At the operational level, German commanders conducted counterstrikes in various stages of Red Army offensive operations, with differing implications for the Red Army commander's reaction and unit capabilities. German counterstrikes were usually preceded by aviation and artillery preparations while their ground forces approached a line of deployment. In many situations German forces struck from the march. In all cases, counterstrikes were executed from a relatively narrow sector and on a solid front.[14] Often, German efforts at the outset of an offensive faced a Soviet Front that usually had not committed its second echelons and reserves to combat. Additionally, forces designated for penetrating into the operational depth had the full support of first echelon artillery and other combat support. Front commanders usually employed close air support, artillery, antitank reserves and engineer troops in sectors of counterstrikes, allowing penetrating units, with their combat and combat service support intact, to continue moving forward for the breakthrough. While a standing mission for second echelons and reserves was countering German reserves, they would be committed to a repelling fight only in necessary situations. With this amount of combat power available, Red Army commanders found it easier to carry out the task of defeating a counterstrike grouping at the beginning of an offensive.[15]

On the other hand, repelling German counterstrikes in the operational depth, particularly in concluding stages of an offensive operation, was always a difficult affair. Repelling the counterstrike successfully in the depth of the German defense was more difficult to control. A number of factors accounted for this difficulty, such as fighting while spread across a broad area, second echelons and reserves committed already, Soviet formations attrited severely, extended with vulnerable flanks, and rear services over stretched.

Support for flanks of assault groups in offensive operations was a main task for army and Front commanders and their staffs. Commanders took decisive measures to quickly alter the balance of forces in their favor and, by immediate actions, hit and destroyed counterstrike groupings. They accomplished this objective by moving-up existing reserves to the threatened sector, as well as by maneuvering men and weapons from other sectors of the front. Development of an offensive in supporting sectors by other Front troops created a threat to German counterstrike unit flanks or rear. With active involvement, Front commanders personally contacted army commanders and established command posts in the threatened sectors for close, direct command and control of the situation.[16]

This successful recipe demonstrated by the 1st Ukrainian Front in the Kiev offensive operation in November 1943, occurred only nine months after Popov's disaster. On 3 November, General Nikolai Vatutin's 1st Ukrainian Front exploded from the Lyutezh bridgehead just north of Kiev. By dawn on 6 November, Kiev was in Red Army hands. Zhitomir fell six days later, and, by 17 November, Red Army forces were 120 kilometers at Korosten. The deep penetrating wedge threatened to separate the German Army Group South from Army Group Center.

Field Marshal von Manstein, Commander of Army Group South, decided to counterstrike. He ordered XLVIII Panzer Corps, under the command of General Hermann Balck, to Fastov. With a large armored force on the flank of the penetration, as done previously against Popov's Mobile Group, von Manstein planned a series of daring counterstrike operations against his old enemy, General Vatutin.

After early success, with encirclement of lead 1st Ukrainian forces at Zhitomir, German panzer units contested Soviet counterattacks at Brusilov. Failing to achieve their major objective, German units were inhibited from further movement for Kiev. German counterstrike forces faced strong Red Army units from the east at Brusilov and with armor forces threatening at Zhitomir.

The Red Army's creation of mobile obstacle detachments to protect flanks with mixed minefields, in this case, erected in small clusters (five-ten mines) and threw them across the direction of advance by German panzers. In repelling Balck's counterstrike in the area of Brusilov, the 3rd Guards Tank Army mobile obstacle detachments blew up thirty-nine German panzers on mines.[17]

With the Red Army's northern flank unidentified, Balck's panzer corps was ordered to attack on the Zhitomir-Rodmyshl front, rolling it up from west to east. The attack began at 0600 hours 6 December, completely surprising and overrunning Russian 60th Army. The 7th Panzer Division reached Malin, thirty-five miles in the Soviet rear. However, by 21 December, General Vatutin's counterattacks were in full swing using forces in the pocketed area. Balck's corps had bitten off more than it could chew. Two days later the panzer corps was forced on the defensive when Red Army forces attacked from the south at Brusilov. Attempting a feat similar to the destruction of Popov's group, Field Marshal von Manstein failed to eliminate a major armor penetration by the Front's mobile group, General Pavel S. Rybalko's 3rd Guards Tank Army. The field marshal's classic model in the context of the whole war on the Eastern Front became an exception rather than the rule.

In major offensive operations of 1944-45, Red Army forces consistently penetrated defensive lines, fragmented defending forces by successfully repelling German counterstrikes. While German military annals are filled with heroic sagas of their soldiers fending off Red Army attacks against incredible odds at the tactical level, German units were being pocketed in tens of thousands at the operational level by converging breakthroughs in major offensive operations. The list of catastrophic defeats for German divisions, corps, and armies is long; e.g., Korsun, Kovel, Bobruisk, Vitebsk, Minsk, Ternopol, Brody, Yassy-Kishinev. Commanders and their staffs evaluated several factors for repelling counterstrikes. Primary among a multitude of factors were the strength of the enemy counterstrike grouping, the nature of its actions, the operational situation and position of advancing troops, as well as terrain features and time available for the command to respond. Vital for success and sustainment of an offensive operation was the army or Front commander's ability to defeat the counterstrike and prosecute the successful development of an offensive into the operational depth.

Soviet military researchers and analysts identified Red Army war experiences three main methods at the operational level, as with the tactical level, for dealing with an enemy counterstrike force.[18] The first method was to shift the attacking force to the defense. Units would dig-in and establish a stable defense, albeit hastily prepared, with deployed antitank elements, established minefields and engineer obstacles, and coordinated artillery fires. The amount of warning time would determine the degree of preparation for defense. They prepared the force, nonetheless, to receive the counterstrike in an advantageous formation, to stop it, and attempt to pin down the German force. Red Army forces would maneuver a reserve or main force for a strike against the German flanks and rear. As with the tactical counterattack, this course of action was assumed if the counterstrike force was larger than the attacking force.

An operational example is the battle to repulse a German counterstrike in the area of Akhtyrka during the Belgorod-Kharkov offensive operation by the Voronezh Front in August 1943. The Germans conducted their counterstrike against the 27th Army's left flank, which was extended over 170 km. In the center and on the right flank, army forces advanced without a continuous front. Along the line Akhtyrka - Moshenk- Oposhnya, the opposing German commander deployed his strongest counterstrike grouping consisting of more than five divisions, SS "Das Reich" Panzer Division, 7th and 11th Panzer Divisions, units of the 19th Panzer Division, and 10th Panzergrenadier Division, and four artillery regiments. This grouping numbered more than 16,000 soldiers and officers, up to 400 tanks and more than 200 guns and mortars.[19] Initially, these forces were opposed by two regiments of the 155th Rifle Division and part of the 5th Guards Tank Brigade. The 4th Guards Tank Corps, part of the 5th Guards Tank Corps, and a rifle regiment of the 147th Rifle Division, located in the second echelon, could also take part in the first day of combat to repulse German forces. During the offensive stage of the operation which was after the Kursk defense, all participating units were substantially short of men and equipment. On the axis of the German counterstrike, 27th Army had 15,000 men, up to 100 tanks and more than 710 guns and mortars.[20] The correlation of tanks in the sector to those in the German counterstrike was more than 4:1 in German's favor. Considering the counterstrike was on a relatively narrow front, German superiority on the main strike axis increased.

At the same time, German commanders concentrated the main forces

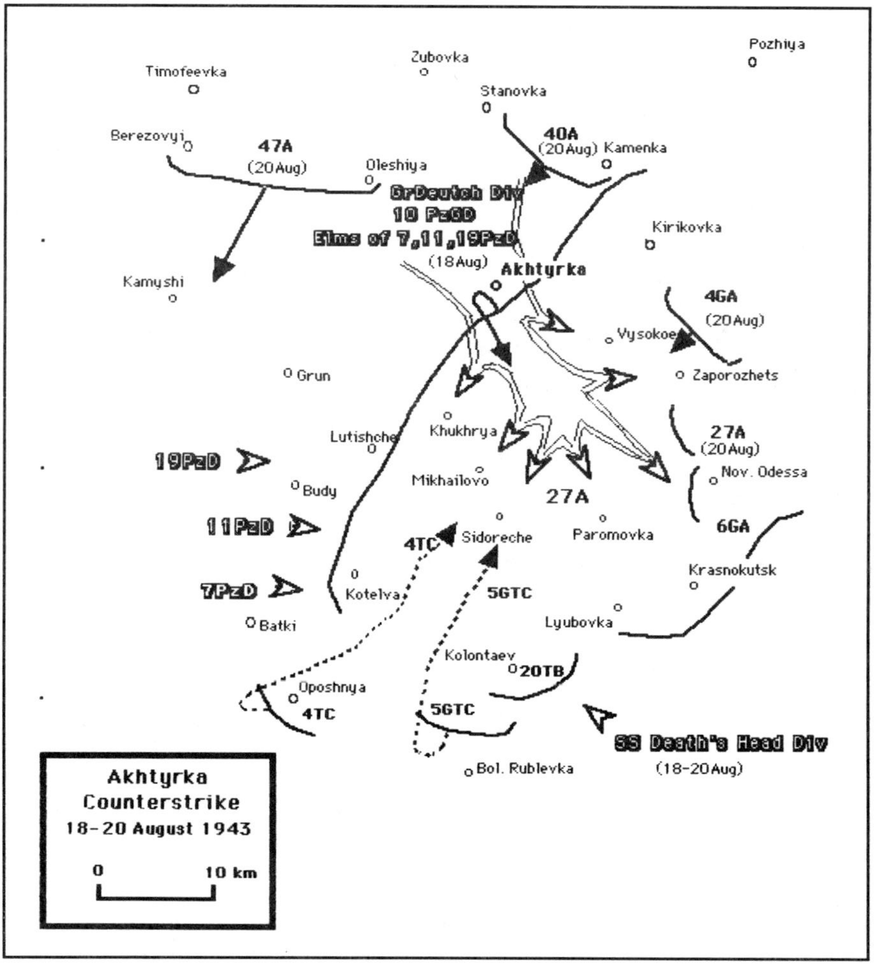

Akhtyrka Counterstrike
18-20 August 1943

of the SS "Totenkopf" Panzer Division in the area east of Rublevka swamp to strike on Kolontayev from the south.

German units began their counterstrike on the Akhtyrka axis on 18 August, following a strong artillery preparation and massive bomber air attacks against Voronezh Front units. For the first half day, the main weight of combat against attacking German panzers fell on units of the 155th and 166th Rifle Divisions.

By simultaneous commitment of up to 200 tanks with motorized infantry, following massive artillery and air strikes, German forces succeeded in breaking through the forward edge of 166th Rifle Division, and, at the cost of huge losses of men and equipment by evening, drove a wedge into 27th Army defense to a depth of 24 kilometers in a narrow 7 kilometers wide sector. In the battle for the forward edge along, the 166th Rifle Division destroyed more than thirty German tanks through artillery fire.

During the first half day, the 7th and 8th Rifle Division, 1st Guards Army, deployed on line Vysokoye-Nova Odessa, entered combat against German units which had broken through.

The same day, SS "Totenkopf" Panzer Division struck from the south. At first, its attack was beaten back by 71st and 241st Rifle Divisions and units of the 5th Guards Tank Corps, defending in this area. Subsequently, German attackers succeeded also on this axis in forcing back large units on 27th Army left flank.

When German forces reached the area of Kaplunovka 27th Army left flank situation became critical, and encirclement of the army's main grouping near Khukhrya-Kotel'va-Kolontayev was threatened. The concentrated counterstrike on the Akhtyrka axis drew the attention not only of Army commanders and staffs, but also of the highest authorities. The Supreme High Commander, Stalin, personally instructed the Stavka representative, G. K. Zhukov, located in the Voronezh Front headquarters, on the need to eliminate quickly the German Akhtyrka grouping, as well as to take urgent steps against possible isolation of 27th Army left flank units.

In order to destroy the German counterstrike, Voronezh Front commander ordered 27th Army with forces of 4th and 5th Guards Tank Corps and 93rd Tank Brigade to strike from the south against the German flank and rear of the grouping operating north of Kotel'va. Additionally, the 147th Rifle Division was to attack from north to south, cutting off the German fighting group which penetrated from its main forces, destroy it, and by evening of 19 August restore the situation. For his part, the 27th Army commander, on the afternoon of 18 August and night of 18-19 August, moved his artillery, antitank reserves, and regrouped an additional one-third of an artillery unit and almost all of his engineer resources to the axis of the German penetration. At the same time engineer units laid minefields. During the fierce meeting engagements, lasting throughout the 19th, Red Army forces stopped further German movement. Air assets from three air armies were directed to hit German units. The Stavka reinforced Voronezh Front with 47th Army which together with 40th Army inflicted a strike from the north into the rear and flank of the German counterstrike. German forces which penetrated the Akhtyrka area and to the north were beaten back from a number of villages, which still further narrowed the neck of the wedge aimed at Bogodukhov from the northwest. By evening 19 August, the German counterstrike at Akhtyrka was already clearly in a crisis situation. In the end, the German counterstrike was successfully repulsed.

During three days of fierce battles, German units suffered major losses. The 27th Army forces alone destroyed 180 tanks, fifty guns, four batteries of six-tube mortars, and almost completely destroyed up to two regiments of motorized infantry. More than thirty aircraft were shot down in air battles and by antiaircraft batteries.[21]

Despite the eventual success of the operation which exemplifies the first method, the 27th Army's actions revealed significant shortcomings in execution of army actions while repulsing German counterstrike. Reconnaissance was poor, which did not permit timely measures to be taken to repulse the counterstrike. The army lacked a strong and highly maneuverable second echelon and reserves, which could have deployed rapidly to the threatened axes. Cover for forces against German air attacks which reigned over the battlefield was extremely organized poorly. Actions among large units taking part in repulsing the counterstrike were inadequately coordinated.

As a second method, the attacking force would attempt to defeat the counterstrike force in a meeting engagement. This course of action at the operational level often resulted from attacking forces being surprised by counterstriking reserves. Red Army after-action reports attributed such a situation to poor reconnaissance. The study of war experiences around Stalingrad noted 5th Tank Army's clash with German tactical reserves. The 1st Tank And 8th Cavalry Corps had an unexpected encounter with units of the 22nd German Panzer Division in the area of Peschanii, engaging it in battle instead of remaining under cover and bypassing the enemy. The fact that tank forces were surprised by the presence of the German reserve force indicated a lack of reconnaissance by the tank army. The tank force lost a day by being drawn into a prolonged fight, allowing the Germans to put their disorganized units in order and bring up reserves from the rear. This delay showed its effect by slowing up the operation as a whole.[22] Red Army commanders in this situation would attempt to suppress the counterstrike with artillery and air attacks, in order to conduct a meeting engagement on ground of their choosing. Soviet commanders and staffs, considering that the Germans generally employed flank counterstrikes, placed second echelons and reserves to respond to anticipated German actions.

A premier example for this method was 4th Tank Army, 1st Ukrainian Front in the Vistula-Oder Offensive Operation in the area of Kielce in January 1945. The meeting engagement usually developed at a critical point in an operation when both sides sought to control the course of action by attacking.

In this engagement, 4th Tank Army, operating with 3rd Guards and 13th Armies, 1st Ukrainian Front, defeated the German XXIV Panzer Corps. The 4th Tank Army, consisting of the 10th Guards Tank Corps and 6th Guards Mechanized Corps, 93rd Independent Tank Brigade, 22nd Self-Propelled Artillery Brigade - totalling 576 tanks, 174 assault guns and 560 guns, was committed on the first day of the operation with the mission of cutting withdrawal routes of German units in the Kielce-Chmielnik area and assisting the combined-arms armies in seizing Kielce.[23]

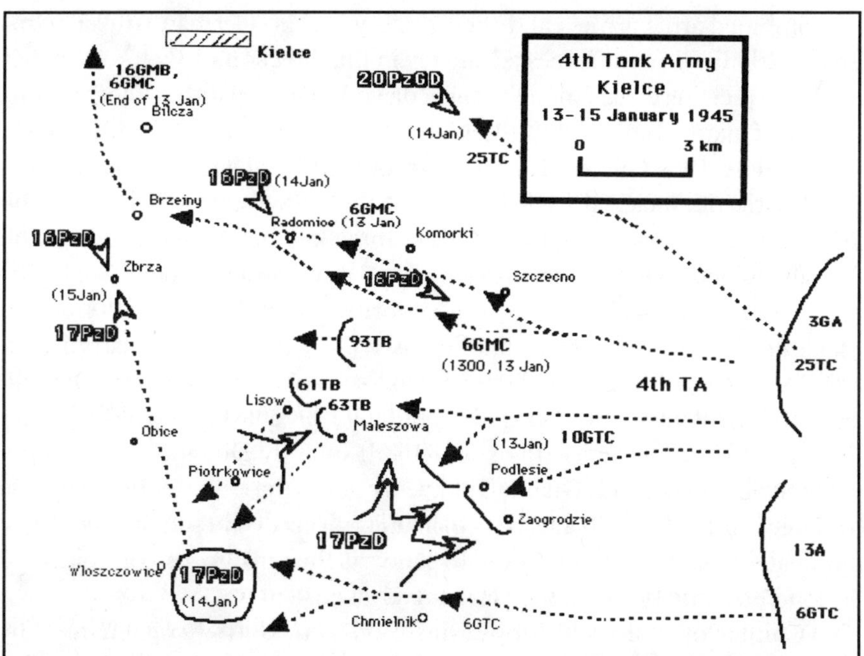

At midday on 12 January, 4th Tank Army committed to the engagement in order to penetrate the German defense and destroy their operational reserve. After combined-arms armies completed a breakthrough of the German's main defensive belt, 4th Tank Army forward detachments penetrated the second defensive belt on 13 January, reaching the Nida river. They were followed closely by tank and mechanized corps which conducted combat operations against the second belt. German forces were withdrawing main defensive belt forces, 68th and 168th Infantry Divisions, to the second defensive belt. During the night, the German command brought up the XXIV Panzer Corps, consisting of 16th and 17th Panzer Divisions, and 20th Panzergrenadier Division - totalling nearly 400 tanks with the mission of a counterstrike against the 4th Tank Army.

The 4th Tank Army operated with the 6th Guards Mechanized and 10th Guards Tank Corps in the first echelon and the 93rd Separate Tank Brigade in reserve. General D. D. Lelyushenko, the tank army commander, planned to continue the offensive in the direction of Podlesie and Zbrza to accomplish his army's assigned mission.

The German command concept of operation sought to hold the second defensive belt by counterstriking with the 16th Panzer and 168th Infantry Divisions to separate the 4th Tank Army from infantry and artillery support of combined-arms armies and then destroy it. The German panzer corps struck 4th Tank Army forces advancing in the direction of Piotrkowice.

The meeting engagement began at dawn 13 January with combat actions by 10th Guards Tank and 6th Guards Mechanized Corps forward detachments. The 10th Guards Tank Corps forward detachment, 63rd Guards Tank Brigade, attacked by lead units of the 17th Panzer Division in the Maleszowa area, was forced to withdraw three kilometers to the south. The brigade joined in a meeting engagement in an unfavorable situation, by deploying its tank battalions into combat formation under intensive German flank fire. The brigade dug in, and repulsed the German attack, carrying out its task successfully. At 0900 hours, main forces of the 17th Panzer Division and 168th Infantry Division units continued the attack. The 17th Panzer Division struck on the left flank and 168th Infantry Division assaulted the right flank, forcing 10th Guards Tank Corps onto the defense. In the course of severe and stubborn battles, German units succeeded in splitting the corps' combat formations. By noon on 13 January, the tank corps committed its second echelon, stopping the German advance.

Combat operations developed differently on the army's right flank. The 6th Guards Mechanized Corps forward detachment, 16th Guards Mechanized Brigade, reaching the Czarna river, became engaged with units of the 16th Panzer Division. During the course of tenacious fighting, German units succeeded in pushing back the forward detachment to Bilcza. But the 16th Panzer Division lost several hours in the process and reached the line of Komorki only at 1300 hours 13 January, where it entered a meeting battle with 6th Guards Mechanized Corps main forces. The mechanized corps, operating energetically and decisively, defeated the 16th Panzer Division in the meeting engagement, and, by evening of 13 January, threw the panzer unit back to the Radomice area. The army reserve, 93rd Independent Tank Brigade, was committed into the gap between the 6th Guards Mechanized and 10th Guards Tank Corps.

Resolute advance by 6th Guards Mechanized Corps and 93rd Separate Tank Brigade threatened the flank and rear of the 168th Infantry and 17th Panzer Divisions, which forced them to halt the attack against 10th Guards Tank Corps units. German units began a withdrawal. The 10th Guards Tank Corps, exploiting success by 6th Guards Mechanized Corps, went on the offensive, coordinating actions with 6th Guards Tank Corps from adjacent forces. By next morning, the tank corps split the 17th Panzer Division and surrounded its main forces in the Wloszczowice area.

During 14 January, the 4th Tank Army, developing its success, forced the Bzhzna river, intercepted the withdrawal routes of the German Kielce-Chmielnik grouping, and defeated units of the XXIVth Panzer Corps and various formations which the German command threw into battle. Combat operations frequently took on the character of meeting engagements. The largest meeting battle developed when 6th Guards Mechanized Corps repulsed 16th Panzer Division's counterattack, pushing the division back to the northwest. During the night of 14-15 January, 17th Panzer Division, with the help of 16th Panzer Division striking from the Brzeiny area, broke out of encirclement. During battles which ensued, German forces were destroyed by 4th Tank and 13th Armies attacks. Fearing further encirclement, German commanders, on 15 January, withdrew units from the vicinity of Kielce.

The meeting engagement, which continued for more than two days, concluded with the total destruction of 16th and 17th Panzer and 168th Infantry Divisions, and a large number of units which the enemy threw into battle. The 4th Tank Army having destroyed a major German counterstriking group in a meeting engagement, came out on the Kielce-Chmielnik grouping withdrawal route. The army mission was completed.

The meeting engagement at Kielce once again confirmed that success was achieved by rapidly seizing the initiative; forestalling the Germans in seizing favorable lines and in deploying and shifting to the offensive; ably exploiting open flanks and gaps to strike the German flank and rear; and quickly splitting enemy groupings into pieces and destroying them. Such a conclusion is indicated by the seizure of an important line on the route of movement of the 16th Panzer Division by 6th Guards Mechanized Corps forward detachment, which made it possible to detain the division for four hours. Forestalling German forces in deploying and shifting to the offensive, 6th Guards Mechanized Corps was able to defeat 16th Panzer Division. The corps advance to the flank of XXIV Panzer Corps main forces created conditions for 10th Guards Tank Corps to switch from defense to offense for splitting XXIV Panzer Corps into parts, encircling and destroying 17th Panzer Division.

The third method, and one that demonstrated the level of Red Army battlefield agility in the latter years of the war, involved shifting part of the main force to defense - containing, pinning and destroying advancing Germans - while the main force would continue to push into the operational depth. This course of action required an availability of forces that allowed containment of counterstrike forces without siphoning-off combat power from attacking forces. The Red Army's major operations in 1944-45 provide numerous examples of this method.

The Lvov-Sandomierz operation was one of the major operations on the Eastern Front in the summer of 1944. In contrast to other strategic offensive operations, a single Front, 1st Ukrainian Front, conducted the offensive with two simultaneous strikes to split the German Army Group Northern Ukraine. The main strike was on the Lvov axis; the second, Rava-Russkaya. These sectors were separated by approximately 80 kilometers.

Demonstrating the third method, the main effort on the Lvov axis dealt with a major counterstrike and local counterattacks against successive commitment of two tank armies to battle in the zone of advance of a single combined-arms army. The breakthrough of the German defense was to be by 60th Army in cooperation with 3rd Guards Tank Army which was to attack on the first day of the operation in the direction of Susov. To the left, 38th Army had the mission with 4th Tank Army to breakthrough the German position.

On the day after the offensive began in the Rava-Russkaya sector, 38th and 60th Armies attacked towards Lvov. Through stiff fighting, lead elements of 60th Army achieved 8-10 kilometers the first day. Encountering battalion and regimental counterattacks, the advance slowed with no significant advance. In 38th Army sector, the situation proved more difficult. General

Hermann Balck's XLVIII Panzer Corps launched a counterstrike with the 1st and 8th Panzer Divisions, pushing back 38th Army's left flank 2-4 kilometers. The two panzer divisions' counterstrike threatened the advancing 60th Army's left flank. Commander, 60th Army, directed his left corps to regroup on the left flank with antitank and mobile obstacle detachments, and he moved his second echelon division to protect against an advance in sector.

The Front commander directed air support against attacking panzers as well as reinforcing 38th Army sector with approximately 400 antitank guns. The Front's aviation support played a major role in repelling the German counterstrike. On 15 July, in the 38th Army sector, 2nd Air Army during the afternoon made around 1,850 aircraft sorties. The air army caught 1st Panzer Division on open roads, inflicting heavy tank losses. Air support helped to halt the German counterstrike.

Meanwhile, 60th Army created a 20 kilometer long, 4-6 kilometer wide breakthrough corridor. A single dirt road ran through the corridor to the village of Koltov which was on 3rd Guards Tank Army's path. The narrow corridor received direct and indirect fires from both sides. Commitment of the tank army's main forces began on the morning of 16 July. German defenders opened with artillery fire on advancing formations and made bomb strikes. Concurrently, General Balck concentrated his panzers and other formations north and south of the penetrating corridor, in order to counterattack at its base, close it, and cut-off the tank army. General Balck created his largest counterattack force in the vicinity of Plugov.

The Red Air Force assigned a sizable force from the 2nd Air Army to support the tank army. The air army made two massed strikes against concentrated German tanks and infantry in the vicinity of Plugov on 16 July. Around 400 airplanes participated in the first strike; 300, in the second.

In response to aggressive actions by Red tankers and aviators, German commanders pulled up fresh reserves from the vicinity of Zagorets, Plugov, and Zolochev. On 17 July, German counterattacks began from the flanks, and by day's end they captured Koltov from the north, Podliptsy and Ivachuv from the south. The stark danger for Red Army forces that German northern and southern groupings would link-up arose. Under these conditions, the 2nd Air Army commander committed more than two-thirds of his forces to cover the tank army. Red pilots attacked German counterattacking groups on the flanks, and they struck strongpoints, centers of resistance, artillery positions, tank concentrations, and approaching reserves. All attempts by German defenders to hinder the Red Army offensive were for naught. The

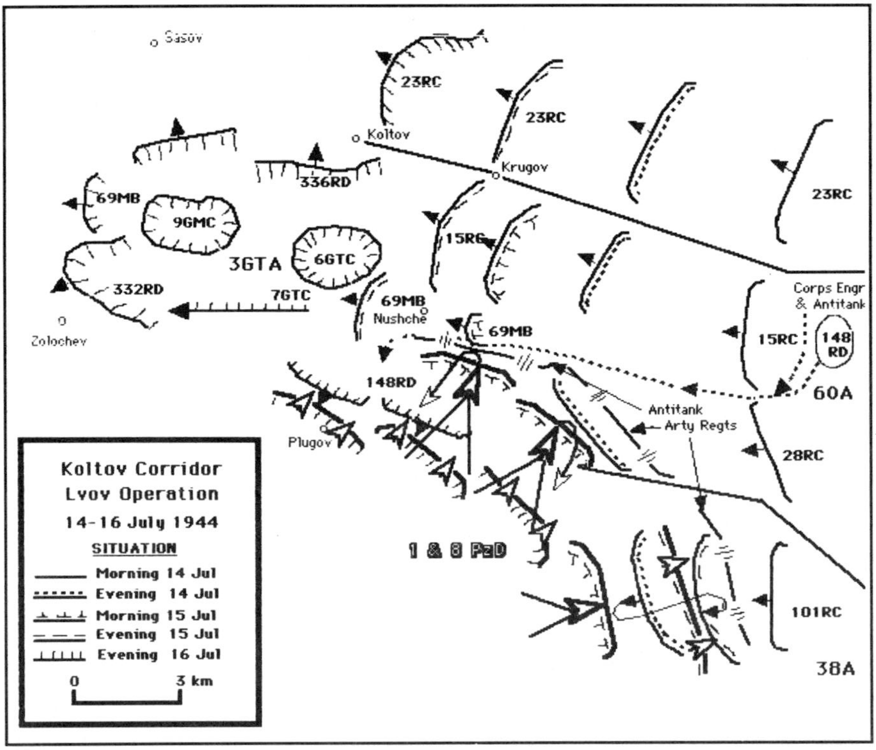

1st Ukrainian Front continued to develop the offensive successfully. An enormous role was played in this by aviation, which flew around 600 sorties just between 16 and 18 July 1944, despite rainy weather.[24]

Attempts by elements of three German divisions to cut off the corridor were unsuccessful. The 3rd Guards Tank Army advanced 30 kilometers on 17 July through the corridor. The 4th Tank Army, which did not obtain a favorable condition for a breakthrough in the stymied 38th Army sector, was directed to move north and into the "Koltov Corridor" after 3rd Guards Tank Army.

Repelling multiple counterattacks, 23rd Rifle Corps, 60th Army pushed ahead on the corridor's northern edge while 15th Rifle Corps fought to extend the corridor. To reinforce the southern edge, 6th Guards Mechanized Corps, 4th Tank Army was used to repel counterattacks. Efforts by adjacent units allowed 3rd Guards Tank Army to attack towards Lvov and attempt to liberate it on the move in coordination with 4th Tank Army. During the 3rd Guards Tank Army's encircling move from the area of Kulikov to the area of Yavorov, the 6th Guards Tank Corps lacked organized flank security. Reconnaissance units traveled only over the main routes, leaving two German ambushes undetected. German forces let the corps main combat force pass and opened fire on rear elements and 7th Guards Tank Corps. However, casualties were light. The 3rd Guards Tank Army used a wide envelopment to reach the rear of German defenders in Lvov.

Conditions for fighting German reserves at the beginning of an operational breakthrough, during exploitation, and at the end of an operation, differed significantly. Conditions for fighting enemy reserves during the time of commitment of armored forces to an engagement were more favorable than in any other phase of an operation. At that time, the armored force was still operating in a compact group, and possessed a second echelon or strong reserve. Its flanks were secured by combined-arms armies, since the tank force had not yet managed to break away from them. Armored troops were routinely covered, in the latter period of the war, by frontal aviation operating from well prepared main airfields. Command and control had not yet been disrupted. The troops had not been in action long enough to suffer significant losses, while materiel reserves had not yet been expended. Red Army armored forces in the operational depth were potent entities for German reserves.

Conditions under which German reserves fought at the time when armored forces were exploiting an operational breakthrough were significantly more complex than at the beginning of an operation. In this period, armored forces usually fought on a broad front, precluding swift concentration of striking power in the needed sectors. Coordination required significant refinements, and sometimes a totally new task organization was required to accomplish newly arising missions. Because tank armies broke away from the main Front forces, they encountered problems protecting their flanks - a function which required allocation of part of their force. In addition, by this time armored unit rear services fell significantly behind, resulting in interrupted troop supply in many cases. These problems were compounded by difficulties in control. The deeper an armored force penetrated German defenses, the more shortcomings arose.

However, conditions of greater complexity evolved for armored forces when they encountered enemy reserves in the concluding phases of offensive operations. By this time, armored forces suffered large losses in manpower and tanks. Reserves of ammunition and fuel were expended to a considerable extent, and because rear services had fallen behind, these reserves, could not be replenished on a regular basis. Coordination, especially with neighboring formations and aviation, became erratic and, sometimes, lost completely, since by this time armored forces operated far from the Front's main forces. Quite often, armored reserves were committed already to the engagement and could not always be promptly restored. Command and control was extremely complex. Armored forces, in the operational depth, fought on a wide front, and the formations carried out different missions, fighting at different depths with large gaps between them devoid of friendly troops. The complexity of the situation in which armored forces were committed against German reserves in the final phases of operation was also the product of the fact that the counterattacking Germans often enjoyed significant superiority in resources, which allowed him to seize the initiative immediately.[25]

These lessons from war year experiences had a great practical value for use in operations and combat preparations. Repelling counterattacks and counterstrikes became an important condition for Red Army forces to achieve their objectives. It became the crucial dimension in transferring tactical success into operational level success. Red Army commanders and staffs had to plan for these contingencies to maintain a continuous operation. They required an indepth foresight, prior planning, and coordinated measures for breaking up counterattack and counterstrike preparations on a short notice. The Soviet Army viewed the Red Army experience as a rich reservoir for research and use in the process of training and educating its forces.

NOTES

1. Paul Carell, *Scorched Earth*, (New York: Ballantine, 1966), p. 210.
2. US Department of the Army Field Manual 100-5, *Operations* (Washington, DC: US Government Printing Office, May 1986), p. 20.
3. Ibid., p. 19.
4. Ibid., p. 20.
5. See A. Evseev, "Otrazhenie kontrudarov krupnykh gruppirovok protivnika v khode frontavykh nastupatel'nykh operatsii Velikoi Otechestvennoi Voinny [Repelling Large Enemy Group Counterstrikes in the Course of Front Offensive Operation of the Great Patriotic War]" *Voyenno-istoricheskii zhurnal* [Military History Journal], hereafter cited as *Vizh*), April 1984, pp. 20-31; S. A. Gladysh, "Obespechenie flangov udarnykh gruppirovok frontov v nastupatel'nykh operatsiyakh [Support For the Flanks of Strike Groupings of Fronts During Offensive Operations]," *Vizh*, June 1985, pp. 10-16; I. N. Venkov, "Otrazhenie kontratak protivnika po opty Velikoi Otechestvennoi Voiny [Repelling Enemy Counterattacks From Experience of the Great Patriotic War]," *Vizh*, October 1986, pp. 29-36.
6. M. M. Kir'yan, "Kontrataka [Counterattack]," *Sovetskaya voyennaya entsiklopediya* [Soviet Military Encyclopedia], hereafter cited as *SVE*, Moscow: Voenizdat, 1977,

4:316.

7. N. G. Popov, "Kontrudar [Counterstrike]," *SVE*, Moscow: Voenizdat, 1977, 4:329.
8. Venkov, p. 29.
9. Evseev, p. 20.
10. Gladysh, p. 10.
11. Ibid., p. 11.
12. A. I. Radzievskii, *Taktika v Boebykh Primerakh [Diviziya]* [Tactics by Combat Example (Division)], Moscow: Voenizdat, 1976, p. 84.
13. V. G. Reznichenko, *Taktika* [Tactics], Moscow: Voenizdat, 1984, p. 103.
14. Evseev, p. 20.
15. Ibid., p. 22.
16. Ibid., p. 28.
17. O. A. Losik, *Stroitel'stvo i boyevoye primeneniye sovetskikh tankovykh voysk v gody velikoy otechestvennoy voyny* [Formation and Combat Use of Soviet Tank Troops During the Tears of the Great Patriotic War], Moscow: Voenizdat, 1979, p. 309.
18. A. I. Radzievskii, *Armeiskie Operatsii [Primery iz opyta Velikoi Otechestvennoi Voiny* [Army Operations (Army Operations From Experience of the Great Patriotic War)], Moscow: Voenizdat, 1977, p. 71.
19. Radzievskii, *Army Operations*, p. 74.
20. Ibid.
21. Ibid., p. 75.
22. Red Army General Staff, *Sbornik materialov izucheniyu opyta voiny*, Vol. 6, April-May 1943, Moscow: Voenizdat, 1943, p. 60.
23. Radzievskii, *Army Operations*, p. 89.
24. I. M. Anan'ev, *Tankovye armii v nastuplenii* [Tank Armies in the Offensive], Moscow: Voenizdat, 1988, p. 405.
25. Ibid., p. 368; Losik, p. 159.

PART III: PERSONALITIES

CHAPTER 9

Nachalnik Razvedki: The Red 'Two'

In the summer of 1943, German forces hoped to resume an offensive drive to victory on the Eastern Front. Red Army troops were fully prepared to repulse that enemy offensive. On 2 July 1943, the Soviet High Command informed the Central and Voronezh Front commanders to expect a German assault against the Kursk salient between July 3 and 6.

General Nikolai Vatutin, Voronezh Front commander, called his army commanders to a late night meeting on 4 July. After the war council, General Mikhail Efimovich Katukov, commander 1st Tank Army, with prompting from his political officer, decided to watch an English comedy film in an improvised field theater. Over the whirl of the projector, a distant thunder was barely audible in the smoke-filled theater. Then, the commander's movie enjoyment was interrupted by a rustling at the back of the tent.

"What have you brought?" Katukov shouted discontentedly.

A low voice told the projector operator to cut the machine and turn on the lights.

"Part of the German forces advance in the attack," Katukov's staff officer reported. "A few have reached the first sector."

"It has begun comrades," Katukov summed. "Everyone to his place." [1]

The staff officer reporting the German attack was the tank army's 'nalchalnik razvedki,' Colonel Aleksei Mikhailovich Sobolev. The nachalnik razvedki, or chief of intelligence, is the equivalent to the U.S. Army S2 or G2 staff position.[2] Aleksei Sobolev began the war as a major, serving as senior assistant to the chief of intelligence for the 33rd Rifle Army. Rising to the rank of Colonel, Sobolev served as the Chief of Intelligence for the 1st Guards Tank Army from April 1943 to May 1945. His length of tour in this prestigious tank formation under a demanding commander and multiple references to him in senior officers' memoirs offer a testimony to his effectiveness.

Combining memoir accounts with other sources on Red Army intelligence provides an appreciation not only of Colonel Sobolev, but also of the critical staff position of an intelligence officer. By understanding their past experience and its empirical basis for development of Soviet Army intelligence on today's battlefield, we will gain some insight into the role of intelligence not only in the Red Army, but also in the present Soviet military command. By specifically reviewing Colonel Sobolev's experience, we broaden our expectations on how a Soviet Army commander may use his chief of intelligence.

In order to survive their war against the invading German Army, the Red Army set up a staff mechanism for capturing lessons learned during the course of the war and passing those lessons to commanders and units. *The Collection of Materials for the Study of War Experiences* written by the Red Army General Staff was a primary publication for dissemination. By the second volume, dated September-October 1942, lessons in intelligence were discussed candidly. The lessons criticized lack of control by commanders and chiefs of intelligence, resulting in deceptive practices by reconnaissance elements. The study offered the following example:

> ". . . in the formations of the 48th Army of the Bryansk Front direction. Having arrived at an enemy wire obstruction,
>
> the scouts often went to ground, and at dawn they returned with the standard report: 'The reconnaissance was conducted without results.'"[3]

The study concluded "the more rapid the commanders of formations and staffs at all levels transfer from paper methods of reconnaissance leadership to direct work on the terrain with those involved, the sooner positive results will be seen."[4]

In line with the study's direct observations, the Red Army 1942 Staff Service Field Regulation defined responsibilities of the chief of intelligence and his staff section at various levels of command. The intelligence section - not surprisingly - was expected to:

- Collect and study information on the enemy, terrain and population in the area of enemy occupation;

• Organize and task all types of reconnaissance and observation efforts according to a Chief of Staff approved plan;

• Prepare intelligence summaries;

• Prepare signal codes for intelligence units in the field;

• Disseminate enemy information within the headquarters, to higher and subordinate headquarters, and adjacent units;

• Study enemy tactics, equipment capabilities and readiness, and maintain enemy battle strength.[5]

The intelligence section maps contained the most detailed and up-to-date information about enemy locations and activities along the front, disposition of his reserves, activities in the rear, and air situation.

Annotations on the map of incoming information were marked by source (see Table 1). The listing of sources illustrates the diversity, redundancy, and capabilities Red Army intelligence officers had at their disposal. While we believe all-source is a contemporary phenomenon, one can see that the Red Army, within the limits of 1940s technologies, had a multidiscipline collection approach, even by modern standards.

TABLE 1: SOURCE LETTER CODE

Source	Code
Prisoner of War	P
Documents	D
Commander's Reconnaissance	KR
Observation	N
Air Reconnaissance	A
Photographic Reconnaissance	F
Infantry Reconnaissance	PR
Cavalry Reconnaissance	K
Artillery Reconnaissance	ART
Artillery Instrument Recon	AIR
Tank	T
Engineer	I
Chemical	KH
Agents	AG
Telephone Tapping	TP
Radio Direction-Finder	RP
Local Population	M

Map posting noted the time of receipt of the various reports and source, for example:

P RP
15.7 10.45 27.10 18.10
(Day, Month) (Hour)

All information requiring verification was accompanied by a question mark.[6]

The regulation recommended all field maps be in code - a practice continued today. For security reasons, conventional map symbols and graphics were not normally used on working maps or papers, particularly if the product was not intended for distribution outside the unit.[7] Cryptic and improvised symbols offer a measure of operations security against the casual observer or loss of working maps, especially important in forward units.

On the Eastern Front, the Red Army used a thin, opaque wax paper for overlays. On wax paper, details of enemy and friendly troops dispositions, and two or three local landmarks were entered in full, or the intersection of grid coordinates, were used to superimpose accurately the wax paper on the map.

For uniformity in designating local landmarks, such as woods, groves, and hills, an orientation diagram was prepared in cooperation with the chiefs of the various sections. Commanders down to company or battery level received the diagram. Designated landmarks provided quick reference points despite the practice of coding map graphics.

The intelligence summary prepared by the Intelligence section, provided, as complete and current as possible, detailed information about the enemy. The summary followed a conventional sequence.

The first part stated general situation and enemy activities to the friendly unit front.

The second part depicted disposition and nature of enemy activities along the front with reference to directions if attacking, from right to left if defending. The source and time of information was indicated. The narrative also identified degree of engineer preparation of positions or limits of advance.

Subsequent parts enumerated spot reports on enemy air force, depth of enemy disposition and reserves, acquired from various collection sources and, again, with the time of receipt.

The last part of the summary stated conclusions on enemy actions, composition, location, and groupings. Also, information which required verification, clarification, or collection was identified, and the means by which it was to be collected were allocated. Upon completion, the intelligence summary was signed by the Chief of Staff, the Military Commissar at the headquarters, and the Chief of Intelligence.

The Chief of Intelligence reported all information to the Chief of Staff and, in certain cases, directly to the commander. As specified in the Field Service Regulation 1942, the staff must prepare for commander "reliable and methodically arranged data" on the situation, with conclusions to enable him to reach a combat decision. "The nature of these particulars submitted to the commander will, in each case, be based on actual fact. Quite often, the commander requires only answers to certain questions."[8]

In December 1943, General Katukov with his chiefs of staff, operations, intelligence, and political officers analyzed the situation around Cherkassy. General Mikhail Alekseevich Shalin, Chief of Staff, was looking over the map.

"Colonel Sobolev," Shalin turned toward the intelligence officer, who was always prepared to report, "read out, please, information about the quantities, technical equipment, and commanders of the units opposite of us."

Colonel Sobolev rustled papers. Katukov, a chain-smoker, pulled out a cigarette and lit it. Colonel Matvei Timofeevich Nikitin, Chief of Operations, spread the operational map on the table.

After reporting, Sobolev sat on the bench; Shalin talked about the German's fighting spirit. Seeking confirmation for his remarks from the intelligence officer, "Discipline is still high, resistance no less. Isn't that so, Comrade Sobolev?" Shalin asked.

"Exactly so," the intelligence officer replied.[9]

Despite the recommendation in the regulation, one can see from this anecdote, at times, the chief of intelligence possessed a subjective assessment of the enemy's fighting quality.

In preparation for the 3rd Mechanized Corps' upcoming battle for Rzhev in November 1942, General David Abrammovich Dragunskii, a well-known Red Army tank commander who served as General Katukov's chief of intelligence, wrote, "General Katukov and Chief of Staff Nikitin in those days did not give me a minute's rest, demanding every new information about the enemy grouping, about their activities and intentions."[10] Aggressiveness and timeliness had already become a hallmark of Red Army intelligence which must feed the information-hungry, empirical process of Soviet command and control.

If sufficient time was available for the preparation of offensive action, reports to the commander ideally furnished the following elements of information about the enemy:

- Place and nature of the action; the extent to which air force and mobile troops are active.
- Strength, composition and disposition in front and on the flanks of the given formation; unit identification; location of command posts; dividing line between units; and their order of battle.
- Availability of reserves and their probable course of action.
- Fighting power and morale.
- Density of the front, the strongest and the most vulnerable points in the battle lines.
- When and where may enemy action be expected.
- New Equipment and tactics used by the enemy.
- Nature of the defense layout. [11]

The commander also required conclusions on the location of the enemy's main force, the destruction of which would ensure victory; any possibility of destroying the enemy piecemeal; the best line of attack to ensure the enemy's defeat; points of stiffest opposition and its nature; and the enemy's probable intentions. [12]

In December 1943, in preparation for the Zhitomir-Berdichev operation, General Katukov held a war council with M. A. Shalin, Chief of Staff, M. T. Nikitin, Chief of Operations, and A. M. Sobolev, Chief of Intelligence, in which he war-gamed on a map the possible enemy courses of action.

"The army moves out on this line. What will 'your' forces do?" Katukov asked Sobolev, who was bending over the map. From Sobolev's first days as chief of intelligence in the tank army, Katukov required him to perform in the enemy's role. In Katukov's memoir he remarked, "At any minute he (Sobolev) must know what 'his' forces intend to do, and give accurate information about the area of their concentration."[13]

"I can move 'my' force from here to here," Sobolev responded, while outlining a circle with a blue pencil (Soviets use blue for enemy and red for friendly), marking the area of German force concentration. He continued, "True, for this I would require three-four days. But, for this period, you are able to move toward Karatin."[14]

"Where would you inflict a strike on our flank?" asked the tank army commander.

"Right here." Sobolev's blue pencil set on the town of Klin.

"They know the forward detachments are advancing here. Here, I would put the covering detachments. What do you think, Mikhail Aleksevich?" Katukov addressed his chief of staff, whose opinion he valued and always included. Shalin agreed with Katukov's decision. The chief of operations quickly marked notations on the map.[15]

Following the decision of the commander, the Chief of Staff organized the staff to plan the battle. The Chief of Intelligence organized reconnaissance and observation and issued the necessary orders for collection. One study of war experience observed "reconnaissance should be purposeful, dynamic, and uninterrupted."[16]

Reconnaissance proved one of the major intelligence activities in a tank army since they always had to be prepared for an encounter with large German panzer units attempting to counterattack a breakthrough. In the Zhitomir-Berdichev operation, the 1st Tank Army fought initially the German's 19th and 25th Panzer Divisions, then the 20th Panzergrenadier Division. During the operation, Sobolev's staff detected the shift of the 6th and 17th panzer Divisions from the adjacent 2nd Ukrainian Front's sector, as well as other units from out of sector.[17]

Early problems in collection planning were lack of fulfillment of these requirements. "Some division and even regimental staffs, instead of working out orders for reconnaissance arising from the actual conditions of the situation, simply reprint orders and even reconnaissance plans sent from above and distribute them to subordinate staffs."[18] Then, the study concluded, that its examples ". . . show, quite obviously, how harmful and even criminal formalism and lack of responsibility are in reconnaissance planning."[19]

The commander would tell his chief of staff what information he required, when he required it, and what forces and means may be used to conduct reconnaissance. Acting on instructions from the chief of staff, the chief of intelligence immediately organized collection with the following considerations:

- Not to disperse the efforts of the reconnaissance groups, but direct them to obtain such information necessary for analytical conclusions.
- Assign additional tasks to reconnaissance units at the proper time during the reconnaissance itself.
- Replace promptly missing reconnaissance troops, or those which

have accomplished their tasks.

• Dispatch fresh reconnaissance units to perform new tasks which could not be performed by reconnaissance units already in the field. For this purpose, there must always be maintained a reconnaissance reserve.[20]

In July-August 1944, the 1st Guards Tank Army penetrating the German defenses near Rava-Russkaya dashed to seize crossings over the Vistula river in the German operational rear area. War experience showed that tank armies required deep, thorough reconnaissance in all directions when preparing to act in the operational depth. During the course of the Lvov-Sandomierz operation, the distance of reconnaissance from forward detachments was 30-35 kilometers, up to 40-45 kilometers from the army's main force.[21]

Katukov showed an additional use and trust for his intelligence officer. He wrote in his memoirs, "Deciding to send forward the 6th Motorcycle Regiment, I assigned the commander the mission: to clear approaches to the Vistula, put across the river in the area of Baranov, capture a bridgehead and hold it until the arrival of the corps forward detachments. Taking into consideration that the regimental commander was still young and inexperienced, I directed Sobolev to help."[22] With a unit of pontoon bridges, Colonel Sobolev moved out with the motorcycle regiment. His unit captured a crossing over the Vistula, and waited for the arrival of the tank army's forward detachments.

On the night of 31 January 1945, in the area of Libenau, a reconnaissance detachment again under the command of Colonel Sobolev discovered a weak place in the German defense and captured a bridge across a canal. The 8th Guards Mechanized Corps followed and developed the tank army's attack towards the Oder river.[23] These anecdotes of Colonel Sobolev's role in reconnaissance illustrate how involved Red Army staff officers became in operations. Early war experience study by the Red Army General Staff concluded, "in those staffs where inexperienced intelligence personnel cannot plan reconnaissance, the Chief of Staff and the commander himself personally are obliged to demonstrate and teach how this is done. The experience of war confirms fully the vitality of the requirement of the field regulation about the fact that the commander himself personally or the Chief of Staff is obligated to assign the order for reconnaissance. It is not at all permissible to transfer the trust of this function to inadequately prepared persons."[24]

Later in the war, when chief of intelligence experience had developed sufficiently, they often moved from the headquarters to execute tasks or get their personal sensing of the battle. For example, just prior to an operation in mid-January 1945, Katukov, together with the Chief of Staff and Sobolev, made their way to the forward detachments. The units were in march column waiting to be committed. Katukov wished them luck, and embraced the lead battalion commander. "What's not said about the first in battle - it's a difficult role. They reach the antitank and panzerfaust fire first, they are the first in mines," Katukov remarked. [25] Sobolev went to the detachment commander to give the latest intelligence for the first in battle.

As the situation shifted from planning to preparation for combat, the Chief of Intelligence required a subtle shift in information emphasis and had to determine:

- What information, in addition to that required by the commander and chief of staff, might possibly be needed for the thorough, accurate appraisal of enemy strength.
- What data on-hand should be checked or confirmed.
- What information can be provided by intelligence units already in the field, by advance units, artillery, adjacent units, and higher headquarters.
- What additional tasks can be assigned to reconnaissance units already in the field.
- What further information fresh reconnaissance units should be dispatched to obtain.
- How to maintain uninterrupted intelligence.[26]

Continuity in conducting intelligence collection can be seen as another hallmark of the Red Army, then, as it continues today in the Soviet Army. Continuity required daily, intense study of the situation. Continuity meant the capability and flexibility to control collection assets, coordinate different types of collection means for complementing collection, and as the Red Army General Staff concluded ". . . only then will there have been achieved the real possibility of protecting troops from anything unexpected in battle."[27]

In contemporary intelligence patois, the collection plan indicated:

- The object of reconnaissance according to the task of the formation, and all available information about the enemy; what information must be obtained, where, and within what time.
- The task of each intelligence collection, the purpose of the intel-

ligence, the direction or area to which collecting should be conducted, the time limits of the intelligence; what the collector is to do after completing its task, in what area prisoners should be captured for interrogation purposes.

• Dispatch of staff officers with reconnaissance units, when necessary. We have seen the intelligence officer himself was subject to dispatch on critical missions or tasks.

• The means and time of delivering reports from reconnaissance units (radio, aircraft, combat vehicles, motorcycles, mounted or foot messengers, message collecting points). All these means are still available on the modern battlefield when communications fail to stretch necessary distances.

In the course of battle or an operation, the collection plan was elaborated and developed. The degree of specification in the collection plan depended on the time available for development. If time was insufficient, the collection plan would be merely notes on a map and in the Chief of Intelligence's personal field book. The collection plan was an operational document of the command post and approved by the Chief of Staff with no circulation outside the headquarters.[28]

One of the most important sources of information about the enemy, according to the 1942 Staff Field Service Regulation and testimonies of Chiefs of Intelligence, was the interrogation of prisoners and deserters.[29] The prisoner was a major source of information for rifle armies which were in constant contact along the front line. Major General M. A. Voloshin, Chief of Intelligence, 39th Army, in his memoirs noted, ". . . prisoners served for us an important source of information. Also with a captured German was usually the soldier's book which had a lot of information. In it was written the soldier's battalion and regimental numbers."[30] The "poisk," or hunt, raids were dedicated reconnaissance teams for capturing enemy individuals or small groups of soldiers in the forward edge of the battle area or close behind it.

This difficult and crucial type of combat reconnaissance were carried out by experienced reconnaissance personnel, called "hunters." Before dispatching a "hunter" group, careful study of the enemy deployment, location of their command and observation posts, headquarters, and communication centers was conducted. Key to a successful mission was knowledge of the routine in the chosen target, such as place and time of sentry shifts, mess hours, and rest time.[31]

The depth of penetration into the German rear area by "hunters" depended on distance to the target and nature of the terrain. As a rule, missions did not exceed 1-15 kilometers while usually 3-6 kilometers. The groups operated behind the German lines several days. The group usually consisted of six-twelve scouts, including two-three sappers. The groups created within the divisions were assigned to the reconnaissance company.[32]

In the Red Army, prisoners gained this way were called "Tongues," and the "hunt" patrol's mission was playfully referred to as "linguistics" - the science and study of tongues.[33] Voloshin also points out that the capture of "tongues," became a highly difficult and dangerous affair. "Reconnaissance personnel not infrequently paid for success with their blood."[34] The "hunt" raids used widely during World War II continue to be emphasized in modern practice. The Soviet Army will seek a "tongue" to verify or clarify their information.

Often times the Chief of Intelligence personally directed the interrogation of prisoners, and many officer prisoners often were taken to the commander. On 15 January 1945, the 1st Guards Tank Army just launching an attack struck towards the river, Pilitsa. "On this day in my command post arrived the first prisoner, a young, tall, smart officer," Katukov recalled in his memoirs.

"Who are you?" Katukov asked through an interpreter.

"Commander, 1st Battalion, 73rd Motorized Regiment, 19th Panzer Division."

"What a division!" Sobolev remarked enthusiastically, unable to contain himself. "How many times it has been defeated, and always continues to wage war!"[35] In addition to Sobolev's presence at the commander's interrogation, his knowledge of German order of battle and a unit's history indicates the degree of the intelligence officer's personal knowledge and continuity of his unit's intelligence holdings.

Documents, correspondence and German publications, captured or found in the field, provided valuable intelligence material. Division and corps headquarters could study in detail only those documents which were of direct practical value to the unit and its particular situation. All documents, especially enemy field orders, topographical maps and overlays, documents of the field coding service and cipher, were immediately forwarded to higher headquarters. [36]

During the first winter of fighting in November 1941, Sobolev, while assigned as assistant chief of intelligence in the 33rd Rifle Army, recalled the capture of a German soldier's diary during a "hunt" patrol. This unknown soldier of the 15th Infantry Division wrote:

"13 November. The Russians shoot all day long, and combat continues at night. From a direct hit in the trench a shell killed seven and wounded ten men. We all have become terrified.

14 November. Heavy artillery and mortar fire day and night.

15 November. Strong artillery fire. Three wounded.

16 November. At 1700 hours a direct hit on the adjacent trench. Killed five and wounded two men.

17 November. Terrifying cold, many frostbitten fingers, hands and legs. One man wounded." [37]

When combat documents about the enemy were received, the Chief of Intelligence would become personally involved with the more important items. They were subjected to careful examination for comparison with information already on hand and with records of previous enemy situations.

The intelligence section also checked the extent to which the new data confirmed suppositions and evaluation regarding the enemy. The reliability of the information was determined. The German documents were appraised for important information on Soviet troops and what the Germans knew about the Soviet situation. All information revealing the actual situation and intention of the enemy was compiled and summarized periodically.[38]

Any further intelligence collection requirements were determined. If any doubts arose as to the reliability of the information received, it was verified immediately, either by dispatching fresh reconnaissance units or by assigning additional tasks to those already in the field.

In December 1943, in fighting around Berdichev, General Katukov remembered an incident in which there was a need to reconfirm an intelligence report. "We had a basis for fear for our right flank. The enemy in the area transferred from the town of Malin two divisions." And, not unusual for a tank army penetrating into the operational depth, Katukov added, "We were uneasy about the left flank. Then more so, when air reconnaissance reported in the woods east of the town Ruzhin, a concentration of approximately 100 enemy tanks. As a rule, this information needed verification. This is why Sobolev sent to this area motorcycle reconnaissance which combed the woods and spotted nothing except 84 precisely formed wood stacks. The pilots mistook them for enemy tanks."[39]

Information received from what we call today human sources, always had to be verified unless previously confirmed by data obtained from other sources.[40]

In reports to the commander concerning the enemy, fact must be clearly distinguished from conjecture. General Dragunskii described General Katukov's words of admonition upon Dragunskii's arrival in the 3rd Mechanized Corps.

"All right, I understand you myself, because, keep in mind, in the past I, too, was an intelligence officer. The wisdoms of this profession are familiar to me. Therefore, do not try to lead me. About the enemy you are obliged to report the truth and only the truth. I like accuracy, and I do not tolerate approximations. Consequently, in no way must your vocabulary have 'obviously,' 'presumably,' 'requires more accuracy.' Throw out these words once and for all."[41] Implicit in Katukov's demand was a reliability achieved through accurate observations and specific analytical conclusions based on facts. No facts without comment, and no comments without facts.

A briefing technique Colonel Sobolev used to establish credibility was a calm, monotonous, enumeration. "On the basis of air reconnaissance information, air observation, visual observation, prisoner interrogation, studying documents, and information from reconnaissance groups running through the operational depth of the enemy's territory, and material, taken from the whole line of destroyed, captured staff . . ." Usually his litany as a preamble forced even the strong-willed commanders to hide an irrepressible desire to yawn.[42]

However, when Sobolev finished his preamble and began to discuss information of interest, the commanders paid close attention. ". . . it can be concluded, that against us will operate an isolated part of the 21st Army Corps with 500 machineguns, 150 guns of different calibers, assault guns, tanks and mortars. The enemy's gun is the panzerfaust: firers in the fortified region and partly situated in fortifications. The full capacity of the fortified region numbers a minimum of six full strength divisions . . ."[43]

During a combat situation, the field regulation recommended reports begin with the most important data and observe the following order: time at which the enemy was discovered; where; by whom; his strength and composition; reliability of the information and brief conclusions as to the condition, activities and probable intentions of the enemy.[44]

The most important intelligence reports were immediately submitted word for word. For example, at the height of the defensive Kursk battle on 7 July 1943, Katukov, anxiously battling German armored forces on the southern face of the salient, recalled: "Sobolev came to me."

"Comrade commander, an intercepted message." He placed before Katukov a sheet of paper. "You see, that is a German air reconnaissance report."

The text read: "Russians are not falling back. They are standing in there on the line. Our tanks have come to a standstill. They are burning."[45]

This timely and important message came at a critical time which allowed Katukov to make the assessment ". . . the Germans did not achieve an appreciable success on the Oboyan direction, they began to be nervous."[46]

In offensive operations, the typical methods of conducting reconnaissance in the 1st Guards Tank Army were bold and decisive actions by reconnaissance groups, separate reconnaissance patrols and dismounted reconnaissance parties, raids against the enemy's withdrawing columns, ambushes along the routes of his withdrawal, and mobile officer observation posts. While widening the bridgehead on the west bank of the Western Bug river in the vicinity of Sokal during the Lvov-Sandomierz operation, a separate reconnaissance patrol of the 8th Guards Mechanized Corps, 1st Guards Tank Army, circumvented German strongpoints and penetrated 18 kilometers into the German defenses, where it organized an ambush. A column of withdrawing motor vehicles was destroyed and prisoners were taken for identification.[47]

When the tank army assumed the defense, Sobolev used raids and ambushes, sending listening sentries to the line of contact with the Germans. Deep actions by dismounted reconnaissance parties dropped in the German rear, officer mobile observation posts, collection of captured documents, and mutual information with rifle armies were the principal collection methods. In holding the Sandomierz bridgehead in August 1944, a dismounted reconnaissance party of the 1st Guards Tank Army consisting of 11 persons was sent to the vicinity of Tsmelyuv. For three days it remained in the German rear, 14 kilometers from the front line, transmitting information by radio on the approach and regroupings of German tanks and infantry, and on the locations of artillery firing positions.[48]

Information about the appearance of fresh German units, their disappearance from places in which they had, or were supposed to have been, and any radical change in enemy activities were immediately reported to the commander and the chief of staff. Such reports were expedited to higher headquarters and communicated to subordinate and adjacent units.

In addition to working maps, the intelligence section produced order of battle charts. When a new enemy unit appeared on the front, the intelligence section organized a systematic study of its condition assessing strengths and capabilities. Corps and division levels kept a "field chart of enemy forces," on which intelligence sections entered all information on the battle composition of enemy troops. Sobolev demonstrated command of this aspect of his intelligence section by his quick response to the 19th Panzer Division officer.

In the offensive approach march, the intelligence section had a set of concerns that required a continuous intelligence collection effort of all sources during the period of the approach. Collection requirements sought to determine with greater accuracy the organization of the defense zone, German unit disposition, and system of defensive fortifications.

Time limits for the preparation of photo-diagrams and for the compilation of data provided by all the collection means was limited. In addition to the main and outer defense zones of the enemy, the intelligence section attempted to develop photo-diagrams on the second (rear) enemy defense line.

Approaching the German defensive zone, intelligence sections worked to elaborate and amplify the collection plan on the enemy's main defense zone and its obstacles. A system for methodical ground observation during the period of combat of the forward detachments was organized.

When the security zone had been overrun and the forward detachments arrived near the enemy's combat outposts, divisional headquarters organized the destruction of these outposts and reconnaissance of the defense zone.

During this period, the main task of all headquarters was the direction of troop reconnaissance and the direct execution of this reconnaissance by staff officers and the chiefs of special army branches.

The work of intelligence was paramount at this stage. It concentrated all its collection efforts on ascertaining the enemy defense system on the forward edge and in depth. On another occasion in January 1945, Colonel Sobolev was called to lead the tank army's reconnaissance - this time on the approaches to Poznan. He conducted a deep reconnaissance seeking to determine the obstacles and barriers of the town's fortifications. The reconnaissance unit under his command captured the German deputy commander of the fortified area. In addition to the prisoner, the Chief of Intelligence returned with a map of Poznan.[49]

As forward detachments engaged the enemy, intelligence organized direct observation of the battlefield along the entire front of the division. The entire zone of advance of the divisions divided into observational sectors. In accordance with gradual progress of advanced guards, the observers proceeded from one observation post to another, in order that they would fully observe the forward edge and, if possible, the depth of the enemy disposition.

The observation system of artillery chiefs and their command posts had to be organized with particular thoroughness. At the same time as reconnaissance of the forward edge was conducted in battle, the observation, reconnaissance, and photography of the enemy defense zone continued from the air. The Chief of Intelligence took all measures to deliver timely information to the commander.

The Red Army Staff Service Regulation identified actions for the Chief of Intelligence and intelligence section in all aspects of combat. In addition to guidance for offensive battle, it identified actions on the march, meeting engagements, defense, mountain warfare, crossing water obstacles, fortified areas, and in winter conditions. Intelligence became an integral part of all aspects of warfare.

In the context of their past experiences, modern Soviet Army battlefield intelligence can be expected to continue to strive for those hallmark principles evidenced in the actions of Colonel Sobolev - reliability, continuity of effort, aggressiveness, timeliness, and focused collection.

Forecasting the impact of technology on the future battlefield, the Soviets believe the acquisition and processing of intelligence will be an increasingly vital dimension of combat. New technologies have increased the range of collection, diversity of information, and consequently, the quantity of intelligence on the battlefield. The intelligence staff position plays a key role in what the Soviets call, the "struggle for information."

The "struggle for information" is the proliferation of the collection systems to acquire the information and link that acquisition to firing systems, such as reconnaissance fire or strike complexes, for quick response. The other half of the struggle is denying accurate information being gathered by the enemy on friendly forces.

In the struggle for information on the future battlefield, we must not underestimate the capabilities of chiefs of intelligence to accurately assess our actions and capabilities. We must be strong advocates for deception and operations security to deny their intelligence collectors information necessary for their quantitative analysis in decision making. We must work with the operational goal of directing the enemy towards friendly plans and intentions. Winning this struggle means affecting the enemy command and control through his intelligence system - we must beat the Red TWO.

Postscript: Sobolev retired immediately after the war. "For a well deserved rest," Nikolai Popel, his army political officer, observed.

NOTES

1. N. K. Popel, *Tanki povernuli na zapad,* Moscow: Voenizdat, 1950, p. 108.
2. The Soviet term "razvedka" translates both as reconnaissance and intelligence. The term includes the western armies' concepts of reconnaissance, battlefield surveillance, target acquisition, collection management, analysis, production, and dissemination.
3. General Staff of the Red Army, *Sbornik materialov po izucheniyu opyta voiny, No. 2,* (hereafter *Sbornik #2*) September-October 1942, Moscow: Voenizdat, 1943, p. 47.
4. Ibid.
5. *Red Army Field Service Regulations, 1942* (hereafter *FSR 42*), translated under the direction of the Chief of General Staff, Canada, July 1944, paragraph 30.
6. Ibid., para 51.
7. *Handbook of Soviet Armed Forces Military Symbols,* Defense Intelligence Agency, DDB-2680-41-78, August 1978, pp. 3-4.
8. *FSR 42,* para 166.
9. Popel, p. 263.
10. D. A. Dragunskii, *Gody v brone,* Moscow: Voenizdat, 1975, p. 87.
11. *FSR 42,* para 167.
12. Ibid.
13. M. E. Katukov, *Na ostrie glavnogo udara,* Moscow: Voenizdat, 1976, p. 273.
14. Ibid.
15. Ibid.
16. *Sbornik #2,* p. 48.
17. I. M. Anan'ev, *Tankovye armii v nastuplenii: Po opytu Velikoi Otechestvennoi voiny 1941-1945gg,* Moscow: Voenizdat, 1988, p. 198.
18. *Sbornik #2,* p. 48.
19. Ibid., p. 49.
20. *FSR 42,* para 191.
21. Anan'ev, p. 202.
22. Katukov, p. 322.
23. A. Kh. Babadzhanyan, et al, *Lyuki otkryli v Berline,* Moscow: Voenizdat, 1973, p. 255.
24. *Sbornik #2,* p. 49.
25. Katukov, p. 345.
26. *FSR 42,* para 192.
27. *Sbornik #2,* p. 59.
28. *FSR 42,* para 194.
29. Ibid., para 199; M. A. Voloshin, *Razvedchiki vsegda vperedi,* Moscow: Voenizdat,

1977, p. 17.
30. Voloshin, p. 17.
31. *Sbornik #2*, p. 54.
32. Ibid.
33. Voloshin, p. 18.
34. Ibid.
35. Katukov, p. 347.
36. *FSR 42*, para 200.
37. *Bitva za Moskvu*, A. M. Sobolev, "Boi pod Naro-Fominskom," Moscow: Moskovskii rabochii, 1985, p. 254.
38. *FSR 42*, para 201.
39. Katukov, p. 283.
40. *FSR 42*, para 202.
41. Dragunskii, p. 85.
42. N. K. Popel, *Vperedi Berlin!*, Moscow: DOSAAF, 1970, p. 223.
43. Ibid.
44. *FSR 42*, para 203.
45. Katukov, p. 226.
46. Ibid.
47. Anan'ev, p. 203.
48. Ibid.
49. Katukov, pp. 357-8.

CHAPTER 10

Popel: The Fighting Commissar

Enthusiasm, conviction, and dedication represented the ideal military commissar - a party conscience present in Red Army combat units. A form of the commissar system existed previously in other armies, such as the chaplains of Oliver Cromwell's New Model Army, "Representatives on Mission" in Revolutionary France, and imperial magistrates for the Russian Army during the Russo-Japanese War. Straddling political and military spheres, these representatives of dedicated faith in the midst of chaotic, revolutionary movements provided stability and efficiency to a powerful, military effort. Since party faithfuls are relatively unknown in modern western armies, it seems fitting to profile a tank army political officer who briefly led in combat. As a military commissar, Nikolai Kirillovich Popel exemplified such a keeper of the communist faith, and his part in the fighting on the Eastern Front during the Second World War is an exemplary role.

Before the Bolsheviks came to power, commissars were placed in military units with the Menshevik government of Alexander Kerensky. Bolsheviks installed their commissars on the eve of the revolution in November 1917, winning the Petrograd garrison from the Kerensky government. The Red Commissars' infamy since the Russian Civil War rested in their power of life and death over soldiers and ex-Tsarist officers who commanded Bolshevik fighting units. From political agitators, energetic Red Army commissars who were filled with a belief in the essential necessity of victory, galvanized the military's will.

With the Red Army's creation in 1918, commissars appeared at all unit levels from company to military district. "Without the army commissars, we should never have had a Red Army," Lenin expressed in 1920. The commissar and members of his political sections were instruments through which the Communist party ensured the political reliability of the army. During the

Russian Civil War, the new Red Army required the services of a large number of ex-Tsarist officers whose authority the party had helped undermine only months before and whose political reliability was suspect by the struggling Bolsheviks. Many commanders, staff officers, and most noncommissioned officers served in the Imperial Russian Army. These officers were called "military specialists" (voenspetsy), whose values and interests were believed to be corrupting to the new regime.

In 1918, Leon Trotsky, Commissar of Defense, authorized a dual command in each unit, consisting of a political commissar and a military leader. The political commissar, given almost unlimited power, could countermand any order given by the military leader. They also routinely submitted reports on commanders to higher authorities. Consequently, commissars, selected for their political conviction, knew little about military training and strong tactical leadership. However, as a result of punishments for "abandoning positions entrusted to them," commissars received clear indication that they would be held to the strictest accountability for their unit's fighting conduct.[1]

Young, politically-motivated commissars in early years provoked an unanticipated challenge to army leadership. Wanting to be more than watchdogs for the party, many enthusiastic commissars viewed themselves as men of action, attracted to the mystical lure of combat. The translation of Marxism into practice for communists fumbling with the basic tenets, as one noted historian wrote, "was a tame pastime in comparison with the 'glory' of the battlefield, and the exercise of authority in the very day life of camp."[2] Many commissars switched vocation during the Civil War from political actions to command duties, becoming future Red Army combat unit commanders who answered the inner urge to exercise power directly, to lead men in combat rather than monitor loyalty and check morale.

During the Soviet-Polish War in 1920, weaknesses in divided command and detracting politics became apparent. Possibly responsible for the Red Army's disaster at Warsaw's doorstep, the festering problem grew as leading Bolsheviks struggled to control Soviet military organization and doctrine. A major consideration in party debates as to the shape the nature of the Red Army centered on maintaining a political control of the army much like the Civil War. Through the 1920s, establishment of one-man command in military units was slow. In 1925, at the suggestion of Mikhail V. Frunze, who had replaced the fallen Leon Trotsky as Defense Commissar, the roles of political commissar and military commander were combined in one-man, for a command filling both political and military requirements. A political assistant was assigned to the commander.

Published guidance in 1926 defined military commissar functions for the execution of party-political work, as well as assuring the training and education of personnel in the spirit of class solidarity and communist enlightenment.[3] The manual emphasized that military commissars represented not only the Soviet government, but also the Communist party in the army. The commissar was responsible for the political-moral state of their units and for securing, through the system of party political influence and guidance, successful execution by commanders in their military and administrative functions.[4]

Initially, commissars were temporarily retained in units in which no oneman was considered qualified to hold the dual role. But, by 1928, the positions of commander and commissar had been combined in all units. The 1929 Field Regulations of the Worker-Peasants Red Army defined clearly the role of the commissar. Both commander and commissar were entirely responsible for the condition and combat capability of their military formation. While the commander was personally responsible for the operational command of the troops, the commissar ensured consideration of the political situation at the time of the commander's adoption of an operational decision and implemented the necessary political measures in support of troops' combat activity.[5] Commissars were to be fully informed about the entire combat activity of his formation and all the commander's plans and intentions.[6]

Political work in preparation for offensive combat operations were directed towards explaining the importance of the offensive, creating a definite political mood, focusing attention on training troops to handle their weapons, and studying combat methods needed in battle.[7]

By the early 1930s, military leaders began to eclipse their political counterparts. As one Red Army observer noted, "In the 1920s, the commissar had dominated the commanders, now, thanks to better education and technical knowledge, the commander began to dominate the commissar."[8] In March 1934, the wrangling over leadership in military units settled with the institution of one-man command.

Political personnel of the armed forces shared with commanders in the improvement of their material situation. Simultaneously with commanders, commissars structured their professional core. The Decree of September 22, 1935, established for them nine graduated ranks, corresponding to those of the commanders. The lowest rung on the hierarchical ladder for the political workers of the army comprised within the commanding personnel was the politruk, political director. It was followed by senior politruk, battalion commissar, regimental commissar, brigade commissar, division commissar, army corps commissar and commissar of the army of the 2nd and 1st rank.

With Stalin's purges in the latter part of the 1930s, the principle of a one-man command was abolished, and the dual command principle reestablished. Military commissars were introduced in all troop units. Although orders were issued in the commander's name, they required the signatures of both commander and commissar.

The new statue required, "The military commissar equally with the commander shall be responsible for the political-moral condition of the unit, for the fulfillment of the military duty, and the execution of military discipline by all personnel of the unit from top to bottom, for preparedness for combat operations, and mobilization, for the condition of the armament and military economy of the unit."[9]

The political commissar system instituted during Stalin's purging of military officers, which lasted until 1939, ensured total loyalty in leadership. The numerical losses in military officers and commissars were staggering. The Red Army lost three out of five Marshals, thirteen out of fifteen Army commanders, fifty-seven out of eighty-five corps commanders were shot, as well as 110 out of 195 division commanders. The leveling scythe of the purge swung as low as brigade commanders, of which only 186 out of 406 survived.

Lev Zukharovich Mekhlis was appointed, based upon his political background and personal loyalty to Stalin, to carry out the purge in his first year as head of the political administration. The political administration was also severely struck by the purge. The heads of all the political sections and most of the members of the military districts were arrested. In addition, all army commissars, twenty-five of twenty-eight corps commissars, and thirty-four of thirty-six brigade commissars were removed. One Soviet observer estimated that by the end of 1938, at least 20,000 political workers in the military had been removed. Popel had to survive the purge to rise the position of brigade commissar.

Only after the Red Army leadership's virtual decapitation, Stalin halted the mass murder and wholesale arrests. Approximately 90 percent of the highest ranking officers and about 75 percent of the field grade officers had been shot or had been relieved of their duties. Junior officers received rapid promotions and captains -even lieutenants - were assigned as division commanders.

Political commissars were given not only equality in authority with the commanders, but also assigned the responsibility of countersigning all military orders and of supervising all disciplinary measures. The prestige and authority of military commanders was curtailed further by allowing any member of the Armed Forces to make complaints against him and by subjecting him to criticism at Party meetings. Democratization was widely encouraged; officers were addressed as "comrade" and wore no insignia of rank. Saluting was optional, and an officer who required the military salute, or who employed an orderly for menial tasks, was accused of "tsarist tendencies."

The political commissar system came under a trial by ordeal during the Russo-Finnish War. Inexperienced young commissars shared responsibility with equally inexperienced commanders. The low quotient for judgment worked to the detriment of Red Army units in combat against the Finns. As one western historian noted in numerous Soviet military press articles relating to events in the war with the Finns, ". . . one gets the impression that the commissars and other members of the political personnel in the army actually took upon themselves responsibilities for military matters in action. One reads of a commissar who had stopped the regimental commander wanting to take a personal part in combat, telling him: "The commander has to direct his unit. I shall go to the firing line and figure out the situation myself." [10]

The Red Army's defeat at the hands of the Finns was in part attributable to the complicated dualism of command and to an intangible degree the inexperience of newly appointed officers who were militarily incompetent. Under pressure from the Commander-in-Chief, Marshal S. K. Timoshenko, the political leadership agreed to modifications in the political administration system. The Presidium of the Supreme Council issued a decree in August 1940, replacing the post of commissar with that of a political deputy, subordinate to the military commander. On 12 August 1940, Timoshenko abolished the commissar system. A unitary command strengthened the Red Army's professional officer's elan. The newly created deputies for political affairs exercised fewer rights than their predecessors.

With the military commander's position restored, politically motivated or petty personal complaints could no longer be lodged against him except through regular military channels, and criticism of him at Party meetings was forbidden. His restored authority ensured that he gave military orders and took disciplinary actions without contradiction or supervision by the political watchdog.

Brigade Commissar Nikolai Kirillovich Popel, who found himself on the eve of the German invasion assigned to the 8th Mechanized Corps in the Ukraine, illustrates the positive role of the military commissar in the Red Army during World War II. By the summer of 1941, nearly 70 per cent of political officers had been in their posts less than one year.[11] Popel, too, had been recently assigned to the forward corps.

Saturday evening, 21 June 1941, Popel went to see a Red Army ensemble perform folk songs and dances. After the concert, he accompanied his corps commander to a traditional dinner in honor of the visiting troupe, lasting until the early morning hours on 22 June. After the dinner party, a tired Popel went home. To unwind and relax, he decided to take a cool shower. However, a corps duty officer telephone call interrupted his shower telling him to report to work. The corps commander had unexpectedly called for his staff.

Upon Popel's arrival along with the commander and staff, the 5th Army's Chief of Staff informed the commander by telephone that German forces all along the frontier had opened artillery fire. But the corps was not to give any acts of provocation; they were not to open fire on German airplanes. The corps received an order to wait.

Later, German airstrikes made it quite clear to General Dmitrii Ivanovich Ryabyshev, 8th Mechanized Corps commander, and his staff that Mother Russia was at war. At 1000 hours, 22 June, the corps received orders from the army operations section to concentrate in woods west of Sambor by the end of the day. Compliance with the order required armored units to march over exposed roads for nearly 70-80 kilometers. While on the march, the corps received additional orders relocating their assembly area to the east of Lvov by 1200 hours 23 June, and instructed the corps commander to be prepared to respond immediately to 5th Army commander's orders.

Popel felt good about the party traditions in the unit. He recalled in his memoirs, "I still hadn't foreseen and could not foresee all the diverse forms of party-political works for the impending battle."[12] From this point on, political workers' personal example acquired special significance.

Meanwhile, political workers at the unit level had many confirmed communists who took their responsibilities seriously. Brigade Commissar Popel, a short, stout, energetic man, did understand his role, and he was prepared to share the hardships and rigors of battle with the 8th Mechanized Corps. While moving to their assembly area along exposed roads, German warplanes discovered the corps' divisions on the march. In addition to airstrikes, confusion on orders compounded problems. Towards 1900 hours, the corps received direction to concentrate in the woods west of Lvov. For whatever reason, the corps failed to receive orders, moving it to previously planned position east of Lvov. The corps had to be turned 180 degrees. The commander, chief of staff, and Popel all moved to overtake the divisions and turn them in the Lvov direction.

New orders directed the corps by end of 24 June to be in Brody, and on the morning of 25 June attack the enemy in the direction of Berestechko. The divisions, again, had to road march 120 kilometers. These modern mechanized warriors had not slept for two days.

In the early confusing days, multiple orders, airstrikes, and lack of information resulted in bizarre personal anecdotes and crises across the front. While driving between units, Popel overtook an odd procession led by a young lieutenant and two Red Armymen armed with rifles escorting a stout man with his hands raised above his head. The prisoner, in military field shirt without tabs, limply dragged his legs.

"Who is he?" Popel queried.

"A spy, comrade brigade commissar, we are taking him to be shot," the young officer in charge replied.

The "spy" turned and said "Nikolai Kirillovich, brother . . ." Towards Popel rushed the chief of corps artillery, Colonel Chistyakov. "He was so filled with anxiety that he was in no condition to talk," Popel recalled.

"He is without documents, no vehicle. He was interested in the howitzer regiment. He had colonel's tabs, but a belly like a bourgeois," the lieutenant explained defensively. Popel vouched for the colonel and had him climb into his vehicle. The bedraggled colonel thanked Popel.[13]

Front orders directed the 8th and 15th Mechanized Corps to attack at 0900 hours, 26 June. With the 8th Mechanized Corps concentrated north of Brody, General Ryabyshev decided to attack along two axes towards Berestechko and Petnev. According to the corps commander's concept, the formation would be in one echelon. The 12th and 34th Tank Divisions made up the attack grouping. The 34th Tank Division received the mission to attack the rear of the German 16th Panzer Division south of Petnev. The 12th Tank Division threatened an attack on Berestechko, and the 7th Mechanized Division had the mission to cover the 12th and 34th Divisions from the west. The situation required an immediate attack which allowed the corps only four hours and divisions one and half hours to organize while the regiments organized on the march. General Ryabyshev and his subordinate commanders with staffs worked in tents which became stuffy. The windows were opened but Popel detected no air movement. Commanders wiped their wet necks with handkerchiefs. Popel remembered busy efforts those days. "On this day, three times coming forward to meetings, I tried to strengthen the soldiers' consciousness through personal morale and the soldiers superiority over the enemy."[14]

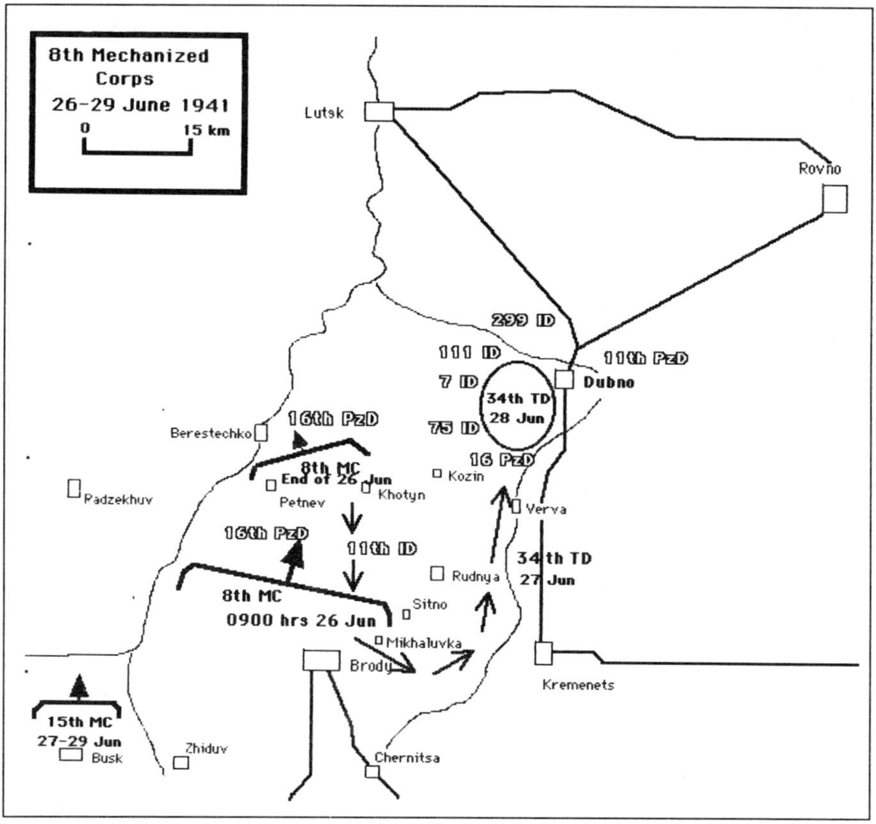

The axis of advance were not a good choice, since it passed through a swampy area of the Slonovka river. In fact, the entire battle area consisted of woods and swamps, the worst for tanks. Four rivers crossed the attack path. Although the rivers were narrow, their banks were swampy and sticky.

Popel and the corps commander moved to the main attack division and put their tanks at the head of the attacking regiment. The operation supposedly had air support. "I glanced upwards . . . not only no air division, not even a regiment or squadron," Popel noted.[15] Without Soviet air cover, the mechanized corps was defenseless to marauding German warplanes. German aircraft used a liquid fuel mixture, like present day napalm, to stop the corps' tanks.

At 0830 hours, 26 June, 12th Tank Division attacked into the 16th Panzer Division flank, and, an hour later, the 34th Tank Division assaulted 11th German Infantry Division. Despite difficulties, the 12th and 34th Tank Divisions inflicted heavy casualties, and, by the end of the day, drove 15 kilometers into the invading German formations. By the end of the next day,

34th Tank Division regiments defeated a German motorized infantry regiment in a meeting engagement, pushing German forces back 30-40 kilometers.

Popel moved with the lead regiment along the road to Leshnev. At a critical bridge crossing covered by German forces, Popel along with the regimental commander exposed their heavy tanks to German fires while other tanks rushed the bridge. Popel's actions gained respect from the fighting commanders and tankers. The day's losses for the corps were twelve tanks and ten more stuck in swamps.

That evening while discussing the day's events, General Ryabyshev, the 8th Mechanized Corps commander, lamented to Popel, "It's an Iron War. Little use of horses remained. My Damascus sword is now just a plaything for my grandson."[16]

By 28 June, German commanders had five divisions opposing the 8th Mechanized Corps, and, with air superiority, they forced the 12th and 34th Tank Divisions onto the defense. Encountering retreating units, Popel could not believe the order to withdraw.[17] Immediately seeking General Ryabyshev, he had difficulty finding the commander or his staff. However, from Brody to the southwest was a single road. Popel drove down it and spotted the corps commander's heavy KV tank parked beside the road. Near the tank Ryabyshev paced back and forth. The general hardly nodded as Popel approached, but took from his breast pocket a folded paper. He handed it to Popel to read. On the page, Popel read the few lines directing the corps to fall back on line with the 37th Rifle Corps which was to the north and on the defensive. And, indeed, the Front commander, General Mikhail Petrovich Kirponos, had signed the order. Such an action consigned the mechanized corps to fight defensively along side a rifle unit rather than use the mobility and shock strength of an armored formation.

Another order received from the Front staff, also signed by Kirponos, directed the corps by morning to attack from Brody to Dubno, and, by evening, seize Dubno. Ryabyshev dumbstruck asked about the previous order. The liaison staff officer who delivered the order knew nothing. By 28 June, confusion came in many forms, not only from foe, but also from friend.

While Ryabyshev and his staff planned to pull together the three isolated groups of the spread corps for the directed attack, the Front political officer, Corps Commissar Nicholas Nicholaevich Vashugin arrived. Vashugin had been a regimental commander until 1938 when he received promotion in the wake of the purges. He advanced to a top political post, Chief Political Commissar, Leningrad Military District. He was transferred to Kiev in February 1941. Described always as invigorated and energetic, Vashugin had already challenged the Front chief of staff's report on the deteriorating situation before General Kirponos. While the chief of staff pointed out factors inhibiting a coordinated response by the Front's units, Vashugin launched a tirade against him claiming the chief had not taken into consideration the "moral factor."[18] Vashugin had now taken it upon himself to go forward and instill the "moral factor."

General Ryabyshev went to report to the thin, short commissar with a big black moustache, "Comrade member Front's Military Soviet . . ."

"For how much do you sell yourself, Judas?" Vashugin accused, cutting him off.

"Won't you listen to me, comrade corps . . ." Ryabyshev attempted to ask.

"You're a traitor," the commissar interrupted, again. "A field trial will listen. Here under the pine tree, we will listen to you, and under the pine tree, we will shoot you."

"Comrade Corps Commissar," Popel moved to intervene in the conversation. They shook hands. "You can require fulfilling the order," Popel continued. They went off to the side to talk.

After half an hour, near the end of their conversation, Vashugin unexpectedly asked, "It does not seem that you are suspicious of the corps commander."

"I was dumbstruck," Popel recalled. "I expected any questions, but not this." Facing each other were the political protectors of the communist party. Each represented opposite roles for the commissar in combat units: one could be an ever present threat to a commander whose mind, already filled with the complexities of battle, did not need the distraction; or, on the other hand, one could be an asset who helped the commander troubleshoot battlefield problems, offering an additional set of eyes and ears for the commander.

Vashugin represented the bad in the system. He intently watched Popel's reaction, "You are quiet. You are aware that he is a former ensign (Tsarist officer)?"

"It is known," Popel responded.

"And that his wife is a Kulak's daughter?"

"It is known," Popel acknowledged, not supporting the commissar's line of reasoning.

Popel could not restrain himself any longer, "You can accuse of whatever you please. However, take the trouble to hear us beforehand."

"And, this is you, a regular advocate for the traitor . . ." the uncontrollably angry commissar rained a flow of abusives on Popel. But Popel would not back down or support the commissar's accusations.

Vashugin walked back over to the corps commander. "Who ordered you to give up territory? Why are you so strung out? General Ryabyshev report."

While the general reported, the corps commissar paced back and forth with his hands behind his back. Vashugin understood things had not gone well, but not to surrender. Looking at his watch, "In twenty minutes report to me about your decision."

The commissar withdrew to his car. General Ryabyshev gathered his staff around him. Twenty minutes was not enough time to organize a reconnaissance, besides adequately considering the requirements to regroup the scattered corps. The corps commander stood up and walked over to the pacing commissar. "The corps can complete its regrouping only by tomorrow morning."

"In twenty minutes a decision, and forward," Vashugin retorted indignantly, in a hardly audible whisper.

"What can we go forward with?" the exasperated corps commander responded.

"I order you immediately to begin the attack," seething, Vashugin in a highly agitated voice, screeched. "If you do not begin, I will dismiss you from your post and put you on trial."

The commissar began dictating an order that his assistant wrote down.

"Give it here," Vashugin demanded, placing the order on the map case. The corps commissar snatched out a fountain pen and with flying ink splashes, signed it.

The order was a suicidal decision. The corps commander, faced with the horrible dilemma of a single alternative, quickly devised a plan. He created a mobile group from the 34th Tank Division, an additional regiment and the motorcycle regiment.

The corps commissar looked sullenly at the corps commander, "When the Motherland wants something of benefit, find a way." The commissar, like many amateurs in warfighting, had much to learn about the limits to running a war on guts of emotion, or the "moral factor." One had to think practically about one's own operations.

Fixing his peaked service cap to his head, Vashugin directed, "Carry it out. And Popel will command the mobile group."

Turning to Popel, the commissar added, "If you take Dubno by evening, you will receive an award. If you don't take it, we will expel you from the party and shoot you."[19]

Corps Commissar Vashugin rushed on to harangue the 15th Mechanized Corps, but the situation there was too far gone for him to do anything. Frustrated, he returned to the Front headquarters. Three days later, General Kirponos ordered Vashugin to conduct a counterattack with a tank division. He led the tanks into a swamp where they became hopelessly stuck and had to be abandoned. He committed suicide.[20]

Popel, in charge of the operation, left for the 34th Tank Division commander's command post for detailed planning. He understood the awesome burden of command in war, "This is not a staff 'game,' not a training exercise, it is the corps' fate."[21] Popel had attended major wargames held in Moscow in January 1941 to determine the utility of mechanized corps and Soviet defense plan. He had an understanding of the potency of armored forces if wielded decisively. Popel's group numbered 217 tanks, most of them the older BT-7 and T-26 light tanks, 25 KV and T-34's, and 9,000 men.[22]

By 1400 hours, 27 June, the plan had been settled, defined, refined and coordinated. Along the main road of advance moved the motorcycle regiment, on the right along a parallel railroad, advanced lead tanks. While moving along the highway, Popel's group overtook rear elements of the German 11th Panzer Division. When lead motorcycles advanced into the German motor column, German drivers and gunners were unable recognize the dust covered Russian cyclists as Red Army soldiers. Once the motorcyclists' machineguns opened fire, confusion reigned. Popel's group caught a tank regiment and rear support units unawares.

Soviet commanders ordered units forward reminding them not to slow the pace; units continued through the night. Another successful battle developed on wide rye fields ten kilometers southwest of Dubno, opening the way into Dubno. Arriving in the town late night, Popel's recapture of Dubno raised spirits in the Southwestern Front headquarters, and recalled the Front operations officer. [23]

In Dubno, Popel's Group could not contact corps. Unable to get information on the situation, his group began preparing defenses in order to hold out until relief came from the corps. Having traveled ten-fifteen hours in their cramped tanks, the tankers had little chance to climb outside and stretch. Despite hunger and fatigue, there was no immediate sleep for the defenders with necessary work to be done.

Colonel I. V. Vasileva, commander 34th Tank Division, clearly fought the mobile group. He technically worked the combat instructions. On a map, he began planning defensive preparations by sector and coordinated fire plans. Popel knew the mobile group was encircled completely in the town. His group could only expect German attacks.

"Our joy in connection with the 8th Mechanized Corps dash to Dubno was premature," the Front's operations officer wrote in post-war memoir. "Dubno became a hornet's nest. And, Popel's group was now locked there in a trap.[24] The Red Army, fighting a withdrawal for survival, could not reverse itself to rescue Popel's group.

After two hours of air and artillery strikes, the German's main effort in the southwest, led by the 44th Infantry Division, attempted to dislodge the weary Russian tankers from the town. As other advancing German units continued to gather around the Dubno bottleneck, the 44th and 225th Infantry Divisions, followed by the 14th Panzer Division, attacked the town from the north. From the southwest, the 11th Infantry and 16th Panzer Divisions joined the encirclement and assaulted Dubno.

Time passed, with still no communications contact. Mounting evidence convinced Popel that the beleaguered mobile group would receive no relief. On 1 July, Popels group repeatedly attempted a breakout from Dubno to the southwest in order to join their corps' main force. However, 16th Panzer Division blocked the way.

The following day, with no communications or any other contact, along with the 8th Mechanized Corps, Popel decided for the group to make its way eastward to a retreating Red Army. "We fulfilled the Front order, took Dubno, held it for six days. Then did not receive new orders," Popel wrote in his memoirs.[25] After six days, Popel, with Vasileva, planned a breakout to the south at 2300 hours. Their intended direction crossed the Ptychi river and passed through the towns of Kozin, Sitno, and Brody, in the direction of the corps' last known location. A column of sixty tanks, each with two rounds for self-defense, prepared for the breakout.

Popel's group eventually broke out with approximately 1,000 soldiers. The venture lost all tanks and vehicles. Colonel I. V. Vasileva perished in the fight.[26] The trek from Dubno behind German lines lasted nearly a month with hiding in woods and small raids against German rear area targets. "Each day was an adventure and problems," Popel recalled.[27] The group moved 650 kilometers behind German lines.

Until Popel's memoirs, the origins of the operation under his command were little known. Marshal G. K. Zhukov in his memoirs remarked, "The operations of the 8th Mechanized Corps could have yielded still greater effect had the corps commander not divided the corps into two groups and handed over command of one group to General N. K. Popel who lacked adequate operations and tactical training to control a large-scale engagement."[28]

Four weeks into the war, reestablishment of a commissar system fashioned by Lev Mekhlis, Army Commissar, resulted directly from low morale and widespread desertions in the Red Army. Playing upon Stalin's lost confidence in his officer corps, Mekhlis restored the humiliating commissar control. A Supreme Soviet decree, "On Reorganization of Political Propaganda Organs and Introduction of the Institute of Military Commissars in the Workers' and Peasants' Red Army," issued on 16 July 1941, assigned commissars to all units. They were given responsibilities equal to those of military commanders. The commissar's authority far exceeded that of line officers of equal rank. "If the regimental commander is the regiment's head, the regimental commissar shall be its father and soul," Stalin said.

Their duties included, according to the Decree of 16 July 1941, "warning the High Command and Government against officers and political functionaries who are unworthy of their position . . . (and) conducting an unceasing struggle against cowards, panic mongers, and deserters." Commissars served in military units below the Army level.

Stalin's watchdogs now threatened the lives of Red Army officers. On 20 July, Stalin ordered all undesirable elements be purged from units. "Panic-mongers," "deserters," "abandoning weapons and battle-position" cost officers and soldiers their lives. On 27 July, the State Defense Committee ordered nine senior officers' death sentence by military tribunal read to all officers and men.

Initial reaction to the measure was extremely adverse. The armed forces considered the decree an insult and an expression of distrust. However, intensified Soviet propaganda emphasizing German acts of atrocity and Germany's colonization plans gradually curtailed desertion and defection to the enemy.

Commissars were to set the example, even in hopeless situations. Hitler's Commissar Order, which directed that upon capture commissars were to be turned over to the Security Service for "special treatment," meaning field execution, only prompted commissars to bitter last ditch resistance for themselves as well as many of their units. Commissars, as all communists and Komsomols, had to give a personal example of self-sacrifice and courage under fire. The regulations stated the party faithful would be examples in action for ". . . rapidly winning the sentiments of the masses and leading them behind the commander at the decisive states of the combat."[29] The most famous picture from the Eastern Front during the war is a commissar leading soldiers in the face of heavy German infantry fire near Voroshilograd on 12 July 1942. Aleksei Yeremenko, the Political Deputy for the 4th Rifle Division, was informed of the division's order to restore the situation in their sector. He shouted in the Operations officer's ear, "We must attack!" He abruptly jumped up and shouted, "Communists, forward! Follow me!" He was mortally shot seconds after the photograph was taken.

Considering the special role of tanks troops, the State Defense Committee resolution, dated 12 August 1941, designated not political instructors, but military commissars, in all tank companies. Political propaganda directorates and departments reorganized into political directorates and departments, which broadened the right of political bodies and increased their role and responsibility for all facets of life and combat activity of the troops.[30]

After rejoining the Red Army in August, Popel was assigned as political officer for a rifle army. Briefly, Popel was with the 21st Army, then he was reunited with his former commander, General Ryabyshev in the 28th Army as the member of the army's military council. The army conducted defensive battles through the winter months until spring of 1942. According to State Defense Committee decree, dated 9 July 1941, "About Members of Army Military Soviets," the Army's Military Council was to consist of three men - commander and two members. In Armies and Fronts existed military councils composed of the commander as official chairman, the chief of staff, and a political functionary who was simply called the "member of the military council." Later in the war when, as a rule, more than one political functionary participated in the council, the senior political man was called "first member of the military council."

Through distribution of responsibilities, the first member of the Soviet engaged himself in operational issues and together with the commander signed many of the operational documents. He led the work of the Political Affairs, controlled military prosecution's office and tribunals. Second member was responsible in distribution control for technical material supporting the force, leadership in sections and rear services. Popel was the first member for the 28th Army.

On the eve of the Kharkov offensive in May 1942, Popel, accompanying General Ryabyshev, visited the forward rifle divisions. On the third day of action, Popel was forward visiting the forward divisions. He reported local victories and defeats by the small units.[31] On 14 May, after the operation began, Popel reported actions of the 6th Guards Tank Brigade, relating the heroics of an encircled tank crew's fight which repelled all attacks.[32] Seeming to take an interest in the tank force actions in the army, Popel visited the 169th Rifle Division's assigned 84th Tank Brigade and its penetration to

Veseloe on the second morning of the offensive.

Fighting of the 28th Army became stalled, while the greater part of the offensive to the south became encircled by a German counteroffensive. The losses of the Kharkov disaster were tolled for senior Red Army commanders. Those claimed: the deputy commander for the Southwestern Front, Lieutenant General F. Ya. Kostenko; commander 6th Army, Lieutenant General A. M. Gorodnyanskii; the member of the military council for the 6th Army, Brigade Commissar A. I. Vlasov; commander, chief of staff and chief of artillery for the 57th Army, Generals K. P. Podlas, A. F. Anisov, F. G. Malyarov; and commander of the army group, General L. V. Bobkin. [33]

By 9 June, the 28th Army was forced on the defensive again as the German Army drove for Stalingrad, deeper in southern Russia. During the fighting, Popel was forward reporting back to the commander on the actions of regiments and visited division command posts.

On 3 July, Popel received an order to report to the staff of the Southwestern Front with further instructions to report to the Stavka.[34]

Recalled to the Main Control Officer in late summer 1942, the conversation with the assignment director began with a question, "What is important to you, rank or work?"

"Work," Popel replied.

Popel received an offer to be the political officer for a newly forming mechanized corps. He accepted the position and met the corps commander, General Mikhail E. Katukov, in the vicinity of Klin, at the edge of Kalinin Misha. In a half-lit barracks, Popel fumbled by the door for the light switch.

"It doesn't work," a voice announced, "there's no light and don't expect any."

"Not only were there no lights," Popel recalled, "but not even chairs and tables. Commanders sat on log blocks, and they worked from macaroni boxes. The next day I was leaving the barracks by the door, when, suddenly, splashing mud, pulls up a Willy jeep. From it leaped a man in a soldier's great coat, on the green tabs were general's stars.

The general walking towards Popel asked, "Popel?"

"Yes."

"Katukov, we will become familiar."[35]

At one end of the barracks was Katukov's room; at the other end, Popel's. Sharing full responsibility often made strange bedfellows within a system of high paranoia. Many of the successful "teams" of political officers and commanders formed a lasting and trusting bond. Popel and Katukov had such a relationship, and stayed together the remainder of the war. But, in the first meeting, in an ageless warrior tradition, Popel would have to prove his worth before receiving acceptance by the fighting corps commander. Despite Popel's representative power, at the sharp end of combat, such things cut little difference in the struggle for survival on the battlefield. Although dual command on the battlefield had been historically catastrophic, Popel seems to have ensured his relationship with Katukov worked. He sought to truly help his combat commanders.

The new command team spent their time preparing the untried corps for combat. The corps engaged in a short action around the Rezhev area in December 1942. In a lull after battle, Popel and Katukov went forward to visit one of the battalions. They sat down with soldiers while they ate. The soldiers were used to seeing their platoon leaders; two generals, appearing in their midst, were unexpected. Understandably, conversation in the beginning was slow, but Katukov and Popel's relaxed manner, cigarettes in hand, and questions like "Well, how are things?" prompted soldiers to start talking. The soldiers typically asked the commander frontline concerns. "When would their battalion get more wheat millet?" "Why is our artillery so stingy with its artillery shells?"

Katukov and Popel handled the questions. After a while, Katukov called for his adjutant who brought a portfolio containing awards. With the soldiers informally gathered around, Katukov presented awards, and Popel read loudly the order, "In the name of the Supreme Soviet Presidium USSR . . ."[36]

Serious Red Army reverses during the spring and summer of 1942 resulted in the abolition of the commissar system just fifteen months after reestablishment. Lev Mekhlis indirectly helped the military leaders struggle against his political watchdogs. Mekhlis had been sent by Stalin as the Stavka representative to the Crimean Front. Under urging from Mekhlis, the Kerch Army Group set out in May 1942 to recapture the Kerch Peninsula from the German Army. A victory would have given the Red Army a strong foothold in the Crimea, presenting a serious threat to a German Army advance in southern Russia and would have afforded significant relief to the besieged Sevastopol. Despite initial Kerch Army Group successes, the Germans regrouped and inflicted a disastrous defeat on the army group. Mekhlis fell into disgrace with Stalin for his blunder while exercising command authority. For his part, he was demoted to the rank of corps commissar.

Senior political commissars' inability to avert the Kharkov disaster, also in May 1942, led to the 9 October 1942 decree on one-man command. The decree entitled "On the Establishment of Full One-Man Command and

Abolition of the Institution of Military Commissars in the Red Army," abolished dual command. The political officer no longer had to countersign every operational order, as was required.

Former commissars possessing the required military competence were given military ranks according to functions they were qualified to perform; others were required to attend military schools before receiving military ranks. Political deputies were subordinated to unit commanders and no longer were permitted to interfere in military matters or to take disciplinary action. They were still required to report on officers, but only through political channels.

Before the 3rd Mechanized Corps could be further tested in battle, Katukov received a telegram. He was to fly immediately to Moscow. He thrust papers into his portfolio. He put on his fur waistcoat and warm clothing - a U-2, bi-plane flight in an open cockpit was not a pleasant journey in winter.

Popel waited with the corps staff three days for his return. On the fourth day, Katukov noisily barged into the hut. His face red from the cold wind, he said:

"So, comrade member of the Army military council."

"Army?" Popel replied, perking up one ear.

"Yes, we're to form the First Tank Army. Lets prepare for serious work."

"I didn't have time to think about the meaning of the unexpected designation," Popel recalled.[37]

The new tank army staff was built on the 29th Army, a rifle army staff. It had its advantages and disadvantages. Katukov was given fifteen days to transfer, form, construct, and then join the Northwestern Front. Upon joining the Front, the newly formed army received ski brigades and an airborne division. The tank army received these odd units to assist in a specific mission, piercing the German blockade around Leningrad across snowy terrain. Key obstacles had to be held by paratroopers for the advancing armor.

In February 1943, Red Army operations began around Demyansk, and, by the end of the month, the 1st Tank Army positioned itself south of these operations for its anticipated use. As the army's forward detachments moved in the attack, new orders directed the army to return for immediate rail loading. The 1st Tank Army was to be transferred, but no one knew where - not even the commander. "We only hear, quickly, quickly, quickly," Popel remembered.[38]

The 1st Tank Army hurriedly loaded onto 100 trains. Popel traveled with General Katukov who kept an eye on his watch and hastily sucked cigarettes. Packs of captured cigarettes filled his pockets to fuel his energy. After passing Moscow, their movement on the rails slowed. They pulled into a station filled with wounded soldiers who told Popel that his unit was going to Kursk.

Suddenly, sirens blared and antiaircraft guns fired. Someone shouted, "Air!"

Bombs fell from the air. A bomb hit between the trains, knocking Popel to the ground. A child's burning felt boot hissed beside him, but he could not see the child.

General Katukov decided to make his way to the staff rail car. Popel headed for the station. Bombs continued to fall. Popel thought of the rail cars with ammunition and fuel. He with other officers climbed into a corner of the station. With night, the trains were moved from the station to nearby sidings.[39]

Early morning fog mixed with smoke protected the trains. Popel drove to the army's assigned assembly area. The tank army spent the next three days off-loading and preparing to assume its battle positions on the southern face of the Kursk salient as part of the Voronezh Front.

Nikita Sergevich Khrushchev, later to become Premier of the Soviet Union after Stalin's death, was first member of the Voronezh Front Military Council. In preparation for the Front's defense on the southern face of the Kursk salient, General N. F. Vatutin's, Front commander, meetings were accompanied by Khrushchev's remarks. In one of the first meetings, after the Front commander made his comments the meeting was given to Khrushchev who drew near him a notebook filled with written pages. Khrushchev reminded the army commanders that German forces were great. It did not pay to make light of that fact nor indulge in superficial bragging. He wanted them to keep their feet on the ground and their heads out of the clouds.

"Ours is a waiting battle from which the course of the war depends very much and the relationship to the wartime allies. Khrushchev opened the strategical perspective for the commanders and their staffs. He also talked some of the practical aspects of the battle. He wanted commanders to insure that they talked to their young soldiers about Tiger tanks and large caliber guns and the inadequacy of their antitank shells. He finished his remarks, "Prepare, comrades, prepare and again prepare! The battle is not far off!"[40].

"The end of June, first days of July were hot," Popel remembered. [41] Generals Vatutin and Khrushchev visited the tank army command post. They warned that the German strike would come against the left and center of the Voronezh Front. The anticipated attack would be against the 6th Guards

Army and the sector reinforced by the 1st Tank Army.

On 4 July, the Voronezh Front commander called his army commanders to a late night meeting. After the meeting, the tank army's political affairs officer talked Katukov and Popel into watching an English comedy film in an improvised field theater. Over the whirl of the projector, a distant thunder was barely audible in the smoke-filled theater, Popel reached over to touch Katukov's arm. Then, their enjoyment in the movie was interrupted by an announcement that the German offensive had begun.

Popel with his commander drove to the command post of the 6th Guards Army. They found the commander, Lieutenant General Chistyakov, finishing his breakfast and sipping tea while giving out instructions. "One could see from the spread on the table," Popel wrote in his memoirs, "that the portly Chistyakov loved to eat, and despite the situation the army commander did not change his habit."[42] Chistyakov, distressed at Katukov and Popel missing breakfast, ordered the orderlies to cover the table again, "Must feed brother tankers." But the meeting was interrupted by artillery fire and the appearance by the 6th Guards Army chief of staff who announced a penetration by a large German force. Popel and Katukov exchanged glances and departed for their units on the front line.

In the 6th Guards Army defense section, Katukov had placed a forward brigade in order to cover the approaches into his second line of defense. The battle lines of the tank brigade were dug in. As the Germans advanced, Katukov and Popel were located well forward in the tank brigade with the company, nicknamed the "Iron Company," since every crew was a veteran. After the company repelled the initial attack, Popel and Katukov drove along the battlefield. The road was pock-marked from German bombings, their Willy jeep bounced.

By midday, it became clear that German panzer units were wedged firmly into the tank army's forward edge. Driving beside armored columns, Popel and Katukov came under air attacks. The German warplanes made regular passes. In front of the command post, the short, thin, General M. A. Shalin, the tank army's chief of staff, stood. His uniform blouse collar buttoned up, his belt tightened in the last hole. Shalin was imperviously calm, "you even with time can be bombed in a crack into which you have climbed."[43]

Shalin shrugged his shoulders as if the air attacks were insignificant. He held more importantly onto an instructing telegram. The new orders directed Katukov to move the tank army forward. Katukov preferred to wait in dugout positions protecting his tanks from advancing German Panther and Tiger tanks. Katukov called the Front commander, asking him to reconsider the order. After consulting with the front military council, Vatutin called back agreeing to the suggestion.

After the first day, Popel, with the political workers of the lower units, summed up the day's results. Some units had suffered serious losses in communist members. Popel visited troops in the hospital, making light conversation with the convalescing and severely wounded soldiers. He continually sought ways to be good company to the fighting man and talked in ways that were approachable for tankers, yet thoughtful. He was gregarious. Pictures of him before gatherings have the soldiers genuinely smiling as if he had just cracked a joke.

The political officer was, or could be, important eyes and ears for commanders. Popel also noted tactical aspects of fighting the new German Panther and Tiger tanks. Moving about the battlefield, Popel operated from the army command post checking staff work and visiting tactical commanders.

On the second day of battle, Khrushchev arrived early morning at the tank army's command post. Katukov was not there. The chief of staff, General Shalin, reported the situation. Then, Khrushchev addressed Popel, "How is it with food? What are you doing about hospitals? Roads? Supplies? Fuel?"

After getting his answers, Khrushchev stayed with the army. His presence did not stifle the headquarters, since he required no one to accompany him. In the course of the day, Popel saw him with the wounded, in the political affair section, in the intelligence section where he talked with a prisoner, and with the chief of staff and the deputy commander for rear services. Khrushchev waited for the Germans with the army. He knew everything that was happening.

After the Voronezh Front defended successfully the southern face against the German armored assaults, the 1st Tank Army was pulled from the line and allowed to regroup for an upcoming counteroffensive. The Front commander, General Vatutin, arrived at the tank army command post, and he announced the tank army and its corps deserved a "guards" designation. He informed the gathered officers that materials and request had been sent to the Defense Commissar.

"In front of Vatutin was a notebook," Popel remembered. "On the first page a column of figures written in violet." In twenty days of battle, the army had lost nearly half of its tanks, and a significant loss in personnel. In the Front commander's figures was an accurate listing of the army's status.

"You hope for reinforcements," Vatutin remarked, reading the thoughts of the commanders and staff. "There will be no reinforcements," he continued. "Not one vehicle, not one man. Such is the Stavka's decision."

"I think," Vatutin continued, "the Stavka cannot, and does not, have the means to do otherwise."[44]

The lightly wounded would have to be returned to the rifle units. Tanks would have to be salvaged from the battlefield. Piecing together the parts from wrecked tanks was apparently the only way the tank army would get its reinforcements.

This offensive operation would be the first for the tank army, and the commanders and staff still had much to learn about large armored warfare. During the course of the Belgorod-Kharkov operation, the tank army ran into difficulties with the brigades that were leading as forward detachments. The problems of several detachments becoming encircled by the German forces drew a visit from Marshal Zhukov.

Zhukov came forward to get the current situation report for major decisions to be made about the operation. Zhukov's indignation was clear. The situation actually turned out tragic for one of the brigades which found itself isolated. As Popel observed in his memoir, Zhukov, spoke "typically" in an overbearing manner in which his unrestrained temper did not increase anybody's will, as if scolding made subordinates work better, move quickly or think things out. But despite this manner, the corps commander bearing the brunt of Zhukov's harsh manner remained decisive and firm, then lost his head.

"Who is in the brigade from the corps?" Zhukov asked.

"My deputy for political affairs, Colonel Serenko," the corps commander replied, apparently rather confused.

Zhukov interrupted him with an attitude as if the political officer had led the brigade into the situation willfully.

"Serenko? The commissar took it in his head . . . to try a tribunal."

Popel could endure no more and stood up for Serenko. He explained the situation.

"Who is this?" Zhukov looked absently at him.

Popel seethed. Zhukov knew perfectly well who Popel was. "We had met ten times or so since the beginning of the war. But now he did not know me," Popel wrote after the war.[45]

The marshal directed abusives towards Popel, but soon cooled. Zhukov disliked political workers. While in Moscow on his way to command forces at Khalkin-Gol in the summer of 1939, his name was on a purge list in an open brief case. He packed his things, collected the necessary papers, and moved to assume his command against the Japanese Army.[46] Now and then, he showed his prejudice and dislike for what he considered their detracting petty tyranny.

After the Belgorod-Kharkov operation, the depleted tank army was placed in reserve for reconstitution. During reserve status, there was much work for the political affairs personnel. Popel oversaw organizing unit Lenin rooms, books, newspapers, magazines, and Army singing and dancing groups for troop entertainment. As Popel paid attention to the actions and works of political workers below him, political officers from the Front advised him.

The political officer was a combination of public information officer, political propagandist, physical training director, special service officer, and welfare officer. The primary duties of the political officer were to supervise and participate in the political indoctrination of his own and lower units; promote and organize recruiting for the Communist Party and the Komsomol; make periodic reports on morale, discipline, and political condition in his own and lower units; arrange for athletic competitions, libraries, propaganda films and posters, newspapers, and other publications for the troops; and act as morale and welfare officer.

While commissars always made much of the well-being of troops, welfare and morale functions of political officers apparently in some cases were performed only superficially. Defectors reported that a political officer, when asked for assistance on a personal problem, usually asked routine questions, promised actions, and later stated that the unfavorable conditions had been overcome. If the soldier, knowing that nothing had changed, disputed the statement, a political argument would ensue, and often the soldier who was involved disappeared. Consequently, the assistance of the political officer was rarely sought.

During World War II, the number of Army officers who were Party members increased, and the officer corps, in general, was considered politically reliable. Consequently, in 1943, the entire organization of the political services was considerably reduced and political instructors at company level were abolished. Former commissars, reduced to political deputies, no longer had the power to control military decisions. Although they had military rank in 1942, the rest of the officer corps neither respected them nor considered them as equals.

J. M. Mackintosh, the British Army observer to the Red Army on the Eastern Front during the war, once remarked, "I recall little political or ideological involvement in the Army or in the minds of officers or men of the Red Army at this time. Political officers went through their routine of ideological indoctrination, but spent as little time as possible on it, and usually restricted themselves to reading out an article or two from 'Pravda' or 'Red

Star', the Army paper. Political officers were tolerated by the fighting men; those political officers who took part in the military activities of their units - learning to use its weapons, digging fox-holes or trenches, taking over effective command in an emergency or helping the wounded, won respect from the men for these activities rather than because of their official role as 'Zampolits,' or Deputy for Political Affairs."[47] This was certainly the role played by Popel, who seemed to have been well received by the military men.

"Certainly nobody feared the political officers," Macintosh added, "that feeling was reserved for the NKVD officers."[48]

Officers and troops alike feared his influence though. To regain some of their former power the political deputies began to draw closer to the OO-NKVD (Army branch of counterintelligence) serving as agents reporting on officers within their units. As informers, they could jeopardize the position of the military commander and exert military control through him, since he was compelled by political expediency to rely on the political deputy for advice and approval. Individual exceptions did exist where the commissar's political position in the party was greater than his rank in the political administration.[49]

During the Berdichev-Zhitomir operation, the 20th Guards Mechanized Brigade fought desperate battles to break out of encirclement. When the brigade broke out of "the circle in a hellish devil's sabbath of bullets and shell fragments," Popel appeared, one of the first to welcome them back behind friendly lines. He greeted the commander as if he had unexpectedly found him in some district committee meeting in the deep rear area instead of on the battlefield. Popel's action prompted the corps commander to remark, "the personal heroism of the party workers certainly led to stimulating and inducing the soldiers into battle for the motherland."[50]

In March 1944, the 1st Tank Army began a major offensive as part of the 1st Ukrainian Front. The Red Army had a chance to cut off the German's First Panzer Army and secure crossings over the Dnestr River, making a significant advance to the west. Towards the end of the operation, Katukov and Popel reflected upon the operation, and Katukov confided in Popel concerns about how the army failed to support the corps. The candor of Katukov's reflection reveals the degree of trust the commander had in Popel.

Popel and Katukov were laying on their backs on the damp, sandy bank of the Dnestr River. Their army's effort to cross the Dnestr had been relatively effortless with regards to the Germans. Either the Germans had too little force there, or they intended to allow the crossing and then make short work of the Russians while straddling the river. It was poor in that they did not know about the enemy, their numbers or intentions. Katukov, laying beside Popel, quietly swore through his teeth.

"It was clear to us," Popel recalled, "that even if the crossing went successfully, it was no credit to the army command. We did not help the forces prepare themselves for the dash across the Dnestr. The well-known 'you cannot foresee everything' is little consolation."

Katukov was not comfortable with the well-known aphorism.

"Brigade commanders must think at the corps scale," Popel said, "and corps commanders at the army. And in reality, did we think before the Dnestr at Front scale?"

Katukov lay silently on the cold sand. He closed his eyes and bit his lip. "We easily lecture the others, but don't always adhere ourselves to our advice. As if advice is good only for subordinates. The pontoon-bridge battalion was stuck at the other end of the world - this, agreed, wasn't our fault. But the units advanced towards the Dnestr, poking around as blind kittens. We owed them the location of the crossing points and the enemy."

Katukov turned on his side to look at Popel. His half-length sheepskin coat had grown dark from laying on the damp sand.

"How long a time will you live to still regret our actions?" Katukov asked.

"It will be no less for me than for you," Popel replied soothingly.

"I am not the better for it."

"I firmly believe that Gorelov and Boiko are able to fulfill the mission without our instructions." Popel changed the course of the conversation, hoping to ease the commander's mind. "We forced the Dnestr, higher command gratefully sent decorations to be given. And ought . . ."

"Well, strictly speaking, what ought to do what? Do we take away the awards? Your food for thought is not very creative. Give it some work this self-criticism of yours. Do you like it?"

"Who likes it," Popel sighed.

"That's enough," Katukov rose, shaking off his coat, "In a principle of reciprocity, we forgive each other's sins and we go our way, so talk with your brother political workers, study it by group."

"In spite of Katukov's playful tone," as Popel characterized a dangerous remark by his commander, "I saw, he keenly endured everything about which we had spoken. He suffered the losses, for the kind which he wished he was not responsible, blunders, which nearly no one notices in the general background of success."[51]

Early summer 1944, Nikita Khrushchev once again arrived in the tank army, but this time not with the army staff, but with the brigades. He talked with the soldiers, held meetings, called attention to the heroics and bravery in battles, and talked about restoring the towns and villages in the Ukraine.

In the Sandomierz bridgehead, the tank army fought in cooperation with Front armies for securing a critical bridgehead over the Vistula River. German commanders worked hard to reduce this vital toe-hold. During coordination for the defensive fighting, Katukov and Popel were riding along to the 18th Army's command post when Katukov unexpectedly began talking about Babadzhanyan, one of the tank brigade commanders.

"I confess, when he first came to us, he didn't catch my fancy," Katukov remarked. "Only in the battle for the Ukraine did I discover him. Hot-tempered, bold, and cunning. But thrifty above any measure? And, he doesn't see this in himself, everything is give and give! Give him heavy artillery. 'What for?' I asked. 'Just in case!' Main idea of this man is without reserve, it is not spent."

Popel never interfered with Katukov when he talked to his commanders. He never played the role of challenging the commander's position. Now coming to the aid of a subordinate commander, Popel offered, "But you yourself are not against a reserve?"

"Yes, I am an army, and he is a brigade!" Kakukov responded.

Popel considered the moment and decided to talk candidly with Katukov.

"Babadzhanyan knows perfectly, that a star given to him has only one right - to new feats. His experience is great; he began the war in the regiment. Now, he has been three years in waging war. How many operations has the army conducted! And, in every one he sat in the brigade. What is a brigade? Slightly more than a regiment. You, Mikhail Efimovich, began the war with a division, and in two years commanded an army. It's time to place Babadzhanyan in a corps."

"You are right," Katukov quickly grasped Popel's reasoning, "When we finish the operation, we will offer a promotion."[52]

After the fighting in the Sandomierz bridgehead, the 1st Tank Army was placed in Stavka reserve before being assigned to the 1st Belorussian Front farther to the north. For the army, there came intervals between the battles. Crews were reduced in combat formations. Commanders studied experiences in previous battles and operations. The political workers cited conclusions of party political works, and made lessons for the future. There was no cessation to the party political work in the units. As Popel wrote, concerning these lulls, "There was still a need to keep kindled a hatred for the fascists, devotion to high ideals, and our combat in the name of mankind's happiness."

After three months stay in the reserves, 1st Guards Tank Army received a directive the second half of November to concentrate north of Lublin as part of the 1st Belorussian Front. While the army marched 400-450 kilometers, Katukov and his staff traveled to the Front command post. Upon arrival, Popel met the first member of the Front's Military Council, General Lieutenant Konstantin Fedorovich Telegrin. Telegrin, who shaved his head in the traditional fashion of the pre-war military style, was interested in all details of life in the tank army, cadre, quality of reinforcements, available transport, provisions and many other items. "My report," Popel recalled, "came out very long and detailed."[53]

At the end of December 1944, the Front held a meeting with army commanders and staff for a general characterization of the situation, and put the finishing touches on planning for the impending Vistula-Oder operation. The general depth of the upcoming operation was 400-500 kilometers, from the Vistula to the Oder River.

During the course of the meeting the situation on the western front was reported. General M. S. Malinin, the Front Chief of Staff, briefed that the Germans had launched a counteroffensive on the western front in the Ardennes.

"Not one unit, nor one rifle did Hitler take from the Soviet-German front. In the west, he has total of seventy-three divisions for general numbers, basically Volkstrom, eleven tank and four motorized. They played a basic role in the counteroffensive in the Ardennes," Malinin reported. "English forces and two American armies are falling back in the direction of the Maas River."[54]

In the impending operation, the 1st and 2nd Belorussian and 1st Ukrainian Fronts would strike the Germans' main forces in Poland, approaching the main objective - Berlin. The 1st Belorussian would strike towards Poznan; the 1st Ukrainian had the mission to destroy enemy forces in the vicinity of Kielce and Radomice, and then develop the offensive in the direction of Breslau.

On 2 January 1945, the 1st Guards Tank Army received a directive to develop the mission for the offensive to Poznan. This stage of the operation was 180-190 kilometers. On 14 January, 1st Belorussian forces burst from their bridgehead across the Vistula River. Surprise was complete with devastating results. By the end of the day, the German defenders had been pushed back by rifle units which had already captured bridgeheads over the Pilitsa River.

By 17 January, the army was on the west side of the Pilitsa River, and directed to head west around Lodz. By the end of January, Katukov was looking for a way to pick his way through the Poznan fortified region.

On 2 February 1945, with its forward detachments closing on the Oder River, the staff of the 1st Guards Tank Army had the 1st Belorussian Front's order. Marshal Zhukov, the Front commander, viewed the situation favorably, in that the Germans were ill-disposed with no concentration of forces for any powerful counterattack. German forces were covering only separate axes of advance, but they had been able to transfer four panzer and five infantry divisions from the west.

The 1st Guards Tank Army staff sitting around the map talked about the situation. At first impression the operation looked good, in a week's time they could be in Berlin. Katukov measured the distance, "70 Kilometers in a straight line - ah, that's nothing! Babadzhanyan's corps has covered more than that in a twenty-four hour period. But the flanks of the Front are open!"

General Andreya Lavrentevich Getman, a corps commanders, moving his hand as a fist into a position behind the tank army's depicted positions on the map, remarked, "If I were in Guderian's position I would strike here. Rokossovskii is taken up with the business of destroying the East Prussian enemy concentration. He has turned his Front to the north and thus the right flank of our Front stretches for about 300 kilometers. Konev has turned his Front to the southwest and his left flank runs for 120 kilometers. What could be a more favorable position for the Germans for mounting a blow!"

General Shalin, putting on his dark, round-rimmed glasses leaned over the map, "That little balcony is full of danger." He traced with a pencil a huge salient on the map from the north on the Oder river to Konigsberg. "Here, there are more than nine German infantry divisions, some tank divisions, and a mass of combat groups. And, the main thing is that they are continuously gathering together new forces. In such a situation as this, an immediate offensive on Berlin would be extremely risky."

"Well, possibly political information demands an immediate seizure of Berlin?" General Katukov asked.

"I know only what the others know," Popel shrugged his shoulders.

A strong sense of concern began to grow in the staff and commanders as they worked deeper the operational concept for the new order. General Getman, again, "The alphabet of war is that victory does not depend on taking towns but on the destruction of enemy forces. Napoleon in 1812 thought otherwise, he lost Moscow, and we all know how that turned out. And, he was no small leader of men in war. Tell me frankly, can we go forward?" Getman addressed Katukov.

"Why yes we can."

"Can we take Berlin?"

"Well that would depend . . ." Katukov attempted to remain realistic about the order.

"Well, what doubts have you got?" Getman continued, "it is necessary to expand out the bridgehead, to bring up the rifle armies, and the main thing is to liquidate the 'Pomeranian Balcony.'"

Getman, caught up in making his case, addressed the group, "The army's tanks require a technical inspection. By direct route they have covered 570 kilometers but, in fact, according to what is registered on odometers, they have covered 1200. A man has no odometer, and nobody knows what wear and tear has taken place there in the last eighteen days without rest and, now, without any kind of preparation. Then rush into the attack shouting "hurrah" and into the most difficult new operation where the enemy will fight to the last SS man, and where on our flanks there remains three undestroyed enemy armies. But, Hitler has one hope that we are going to lay our head into his noose and then from the north, from Pomerania one blow, and then from the south from Glogow two blows. He strikes and destroys!"

The war council stood silent for some time. The telephone rang. The Front's member of the Military Council, General Telegin, asked, "Comrade Popel? have you received the order and the instructions?"

"Exactly so," Popel responded, in the Red Army equivalent to "Yes, sir."

"Have you gone down to the troops?"

"Well, no not yet," Popel replied cautiously. "We have just been sitting; we have been turning things over in our minds."

"Then, hang on a while," Telegin directed, "that is a personal instruction from Zhukov. They are deciding things now in Moscow. So wait for a new order."

All eyes of the group were on Popel. He announced to the group, "We have been ordered to hang on. We are not the only clever ones to be found. Andreya Lavrentevich (Getman) rather lost his nerve for nothing."

"Don't waste your sympathy," Getman replied.[55]

In March, the 1st Guards Tank Army was directed north to help the 2nd Belorussian Front reduce the Pomeranian Balcony. The operation began on 1 March. Only at the end of March did the tank army's offensive action conclude since beginning in the Magnushev bridgehead on the Vistula River, 15 January.

After successful operations of the tank army with Rokossovskii's 2nd

Belorussian Front in the Pomeranian operation, the Berlin offensive became the main focus. In the suburbs of the small town of Birnbaum, in a building which had been used as a club, Marshal Zhukov, commander 1st Belorussian Front, gathered the army commanders, members of the military councils, commanders of independent tank and cavalry corps, and other commanders of formations for a briefing on the Berlin operation.

Marshal Zhukov entered, and behind him followed General Telegin and General Malinin, Front chief of staff. Looking over the assembled group, Zhukov began, "Comrades, I have been with the Supreme Commander (Stalin). The situation is such that I thought it necessary to summon you here urgently. Earlier we had suggested that the Berlin Operation would begin somewhat later. Zhukov coughed to clear his throat and continued, "Now, the timetable has been changed. Our little allies have been hurrying us along with requests that are not entirely allied. Eisenhower intends in April to surround and destroy the Ruhr group of the enemy, and then to move on to Leipzig and Dresden and 'incidently' take Berlin. All that will be carried out under the aspect of helping the Red Army. But, the Stavka is very well-informed of the fact that in truth taking Berlin before the arrival of Soviet troops amounts to the same intention of Eisenhower. The Stavka is also well-informed that they have hurriedly prepared two parachute divisions which will be dropped on Berlin. The Hitelerites, as is obvious, are arranging this. They are offering us stubborn resistance in each occupied population center, but on the Western Front even by telephone they are surrendering big cities."

Zhukov continued, "All this has inclined the Stavka to press on with its plans. Marshal Rokossovskii is to destroy the East Prussia concentration of the enemy; in the forthcoming week we will assign him the right hand sector. Rokossovskii has been ordered during the period of one week to move the entire 2nd Belorussian Front 400 kilometers to the west. We have no time to lose and time presses him as it also presses us. The distance to Berlin is not great. It amounts to 70 kilometers. We have done more than that in the first days of the present operation. And arising from this it is impossible to plan to reach Berlin in the first day. But, these 70 kilometers are going to be a hard difficult path. And, therefore, study in advance and with attention to terrain and towns in which we will be committed to operating. It also arises that Katukov's and Bogdanov's tank armies we will plan to introduce on the second day of the operation, and they should be in the suburbs of Berlin towards the evening of the third day! That means exact timing for beginning the offensive and of this I will speak to you later. But, mark well, we will be taking the offensive shortly."

These words livened the gathering. Bogdanov, commander of the 2nd Guards Tank Army shouted, "That's a fine bit of work!"

General Vasilli Chuikov, commander, 8th Guards Rifle Army, added, "The keys to Berlin are to be handed to the tank crews. Well, how do you feel about it tankers?"

"What about us. We are always ready but the main thing will be the infantry securing the breakthrough?" Katukov replied.

"We'll be there, don't worry. It's not the first time for us," Chuikov replied, smiling.

"Now, let us get down to an examination of the tasks which lie before us," Zhukov interrupted the exchanges.

A screen was lifted, and there before the assembled was spread a map. The officers could see it flecked with the spots of the lakes and the winding conduits of the irrigation system. It was plain to see the three defensive zones: the first which occupied the space of land from the Oder to the Seelow Heights and the other two which wound with intervals from 10 to 20 kilometers. Raising a second screen, there came to light a colossal scale relief map of Berlin. With the greatest care there were outlined the streets, fortifications, obstructions, fire points, and even the ruins of buildings which had been bombed. There was the city - laid out. There on the relief the most important government buildings were indicated with little tags which showed the number, they had a number on them of the given objectives.

Marshal Zhukov spoke, "Turn your attention to objective #105. That is the Reichstag. And who will be the first to get there? Katukov? or Chuikov? It might be Bogdanov or Berzarin (commander 5th Shock Army)? And who will be the first to raise the banner of victory?" In a similar manner Zhukov worked through a number of key and important buildings in the city.

On 16 April, the first day of the operation, General Shalin reported the attack bogged down before the Seelow Heights. Katukov unbuttoned his tunic collar and drew in a deep breath. He lamented the stubborn resistance of the German troops. He headed to Dremov's corps, and Popel went to Getman's. Upon his arrival at corps headquarters, Popel met a number of brigade commanders all gathered in conference. "What's holding you up?" he asked. Seelow Heights had proved not to be tank country. German anti-aircraft guns were firing over open sights directly at the tanks, and almost every irrigation ditch impeded their progress. The fighting was getting nowhere. Popel was summoned back to the army staff headquarters.

The 1st Guards Tank Army staff was located in a small dark home is the northern suburbs of Seelow itself. Around a table was sitting General Shalin and on the other side of the table sat Katukov with a telephone receiver in his hand. His head seemed to be resting on his chest, and from time to time he would make remarks like, "Yes, I receive your message - understood."

One of the signals which came through from Marshal Zhukov, "Where is Popel?"

"Out with the troops," Katukov replied.

Shalin turned to Popel, "Kirillovich, the Marshal is calling for you."

It took Popel twenty minutes to get back through to Zhukov.[56]

According to Zhukov's information at this point, the 11th Tank Corps was about 20 kilometers to the west of Seelow. At the end of the conversation, Zhukov gave Popel a personal order: he was to go at once to Yushchuk's (11th Tank Corps) corps, and there organize the redeployment of the entire army.

First, Popel had to find out Yushchuk's actual location. General Popel did not want to ring Marshal Zhukov once again and ask him where to find the corps. So he contacted the commander of the corps' forward brigade, who connected Popel with the corps commander, General Yushchuk, by radio. That night Popel was able to report that he had fulfilled his assignment. Then, he made his way back to the 1st Guards Tank Army's command post.

Late night on 20 April, the army staff received the following message:

"Katukov, Popel.

"1st Guards Tank Army is entrusted with the historic mission to be the first to break into Berlin and raise the victory banner. We entrust you personally to organize the execution. You are to dispatch from each corps its best brigade to Berlin and assign them the mission at any cost to penetrate the outskirts of Berlin not later than 0400 hours the morning of 21 April. Zhukov, Telegin."[57]

A similar message was sent to the 2nd Tank Army. In response to the Front commander's instructions, the 1st and 44th Tank Brigades were dispatched to aggressively move towards Berlin.

At 2400 hours on 24 April, the commander of the tank army's motorcycle regiment sent the following message: "Have reached suburbs of Teltov. On canal have met tank units of Rybalko (commander, 3rd Guards Tank Army)." Shalin had received this report from other units. "Excellent, the 1st Ukrainian Front has reached Berlin," he said. This information was immediately transmitted to Marshal Zhukov.

"Is this report reliable?" Marshal Zhukov asked with a certain amount of disbelief.

"We are reporting on the basis of the combat report of the regimental commander," General Shalin replied.

Zhukov then sent the following signal: "Members of the Military Council will personally go to the forward detachments and ascertain who, in fact, actually first reached the line of the Teltov Canal."

In compliance with Zhukov's instructions, Generals Popel, Getman, and Colonel Sobolev went to the front line, and met a 3rd Guards Tank Army corps commander who took them to Colonel General Pavel Semenovich Rybalko.

Rybalko took them to his observation point which was situated on the roof of the six-story building, and there he showed them the layout of the Teltov Canal. "And this is only one piece of good news for you. I have been talking to General Lelyushenko. He has just reached Potsdam. There he has managed to link up with the forward units of the 61st Army and Polish Army," Rybalko announced.

Popel, remembering his mission from Zhukov, asked, "Well, how long have you been here?"

"We got to the Teltov Canal yesterday evening," Rybalko replied.

Popel made his report to Marshal Zhukov.

"I have a personal order that you were to take the airfield. Have you done this," Zhukov asked.

"Certainly."

"And you took pilots prisoners?"

"Certainly, they've been taken prisoner Comrade Marshal. Operations are proceeding," Popel answered.

"Then, this means that we were the first there?!" Zhukov claimed.

"Comrade Marshal, in truth, those were only small detachments but Rybalko had been there on the 23rd of April. He had broken through on a broad front of some 10 kilometers and had seized bridgehead on the northern bank of the Teltov Canal," Popel did not want to shade the truth of the report.[58]

As the 1st Guards Tank Army moved into Berlin the army had the objective to capture the Reich's Chancellery. On the morning of 1 May, Popel called Telegin and reported that the army had captured the Zoological Gardens and the army was located 200 meters from the Chancellery. On the radio Moscow held its May Day parade in Red Square, Popel recalled, "And, we were located in the enemy's capital, 1000 kilometers from the Motherland."[59] On 2 May 1945, the 1st GTA took nearly 15,000 prisoners. Berlin had capitulated.

In a few days the war was over. Popel had been through many trials on the path to victory. He had built a solid trusting bond with his tank army commander. He fought the enemy or supported his commander. Nikolai Popel exemplified for western observers the positive in the little understood system of the military commissar.

Postscript: As a prolific writer in the post war period, Popel would pen many articles and write three volumes to cover his war memoirs. He wrote a type of memoirs one would expect in the west -a personal account with highlights on personalities. His final volume came under fire from an official book reviewer. Accused of being too subjective, he was criticized for claiming too many personal actions. Popel seems in his declining years to have been isolated politically. He received little mention in other memoirs. His estrangement was a cruel fate for such a gregarious man.

NOTES:

1. D. Fedotoff White, *The Growth of the Red Army*, Princeton: Princeton University Press, 1944, p. 86.
2. Ibid., p. 89.
3. Ibid., p. 233.
4. Ibid.
5. *Red Army Field Regulation, 1929*, (Hereafter *FR '29*) Moscow: Voenizdat, 1929, paragraph 28.
6. Ibid., para 18.
7. Ibid., para 76.
8. Michael J. Deane, *Political Control of the Soviet Armed Forces*, New York: Crane, Russak & Company, 1977, p. 39.
9. Ibid., p. 42.
10. White, p. 402.
11. Dmitri Volkogonov, *Stalin: Triumph and Tragedy*, New York: Grove Weidenfeld, 1991, p. 369.
12. Nikolai K. Popel, *V tyazhkuyu poru*, Moscow: Voenizdat, 1959, p. 28.
13. Ibid., p. 53.
14. Ibid., pp. 63-64.
15. Ibid., p. 75.
16. Ibid., p. 90.
17. Ibid., pp. 129-130.
18. I Kh. Bagramyan, *Tak nachinalac' voina*, Moscow: Voenizdat, 1971, p. 116.

19. Popel, pp. 140-141; D. I. Ryabyshev, *Pervyi god voiny*, Moscow: Voenizdat, 1990, pp. 34-35.
20. Bagramyan, p. 152.
21. Popel, p. 142.
22. D. Ryabyshev, "Ob uchastii 8-go mekhanizirovannogo korpusa v kontrudare Yu-Zapadnogo fronta (iyun' 1941 g.)," *Voenno-istoricheskii zhurnal 6-1978* (hereafter *Vizh 6-78*), p. 71.
23. Bagramyan, p. 148.
24. Ibid., p. 149.
25. Popel, p. 204.
26. Ryabyshev, *Vizh 6-78*, p. 73.
27. Popel, p. 241.
28. G. Zhukov, *Reminiscences and Reflections, Vol 1*, Moscow: Progress, 1985, p. 295.
29. *FS '29*, para 76.
30. O. A. Losik, *Stroitel'stvo i boyevoye primeneniye sovetskikh tankovykh voysk v gody velikoy otechestvennoy voyny*, Moscow: Voenizdat, 1979, p. 340.
31. Ryabyshev, *Pervyi*, p. 195.
32. Ibid.
33. Ibid., p. 218.
34. Ibid., p. 252.
35. Popel, p. 334.
36. Nikolai K. Popel, *Tanki pobernuli na zapad*, Moscow: Voenizdat, 1960, p. 37.
37. Ibid., p. 51; M. E. Katukov, *Na ostrie glavnogo udara*, Moscow: Voenizdat, 1976, p. 191.
38. Popel, *Zapad*, p. 74.
39. Ibid., pp. 77-79.
40. Ibid., p. 105.
41. Ibid., p. 106.
42. Ibid., p. 109.
43. Ibid., p. 115.
44. Ibid., p. 156.
45. Ibid., pp. 164-165.
46. Conversations with Otto Chaney, Zhukov biographer, 5 August 1991.
47. Malcolm Mackintosh, "A Portrait of the Soviet Soldier in the Great Patriotic War (1941-5)," test of an address delivered at the Twelfth Military History Symposium United States Air Force Academy, Transformation in Russian and Soviet Military History, 1-3 October 1986.
48. Ibid.

49. Deane, p. 51.
50. A. Kh. Babadzhanyan, *Dorogi pobedy,* Moscow: Voenizdat, 1975, p. 172.
51. Popel, *Zapad,* pp. 296-297.
52. Nikolai K. Popel, *Vperedi-Berlin!,* Moscow: DOSAAF, 1970, p. 69.
53. Ibid., p. 134.
54. Ibid., pp. 140-141.
55. Ibid., pp. 269-270.
56. Ibid., p. 305.
57. Ibid., p. 315.
58. Ibid., pp. 325-328.
59. Ibid., p. 359.

CHAPTER 11

Radzievskii: The Thinking Warrior

The consummate staff officer, Aleksei Ivanovich Radzievskii, as Chief of Staff of the 2nd Tank Army during the Second World War, applied his keen analytical mind to operational analysis during the course of combat operations on the Eastern Front. He thoroughly understood how to fight the tank army in mobile armored warfare at tactical and operational levels of war. He replaced easily his wounded commander at the helm of the tank army during a major offensive operation in the summer of 1944.

Born on 13 August 1911, he joined the Red Army at the age of eighteen. In 1931, he became a member of the Communist party and completed cavalry school. He commanded a cavalry platoon and squadron, rising to be assistant chief of staff for 1st Regiment, 18th Mountain-Cavalry Division. He graduated from the Frunze Military Academy in 1938 and General Staff Military Academy in 1941.

At the start of the war, Major Radzievskii was Chief of Staff for the 53rd Cossack Cavalry Division from Central Asia. His cavalry unit arrived at the fighting front in July 1941, and unloaded from trains to assemble in woods 45 kilometers to the east of Velikii Luki. The Red Army was slowing the German Army advance with fresh troops from the interior.

In early August 1941, German forces reached the Mezha River line, halted, and began organizing their forces for a further advance towards Moscow. On 11 August, during the lull, the Western Front commander assigned Colonel Dovator's cavalry group, consisting of the 50th and 53rd Cavalry Divisions, the mission to penetrate German defensive positions and conduct raids throughout their rear area near Demidov and Dukhovshchin, northwest of Smolensk. The cavalry units were to paralyze German communications, destroy advancing reserve columns, staffs, and depots, and disrupt lines of communications. The cavalry group, consisting of 3000 cavalrymen armed with medium and light machine guns, operated without artillery or trains.

Until mid-August, cavalry units fought primarily small German units. In preparation for the upcoming rear area raid, Red cavalrymen hid in the forest from German reconnaissance airplanes. From their secluded place, they mounted reconnaissance patrols against German defensive positions.

Cavalry reconnaissance determined that the sector Uste, Podviaze, was occupied by only two battalions of the 450th German Infantry Regiment reinforced with artillery, offering a number of advantages for penetration. Colonel Dovator, at dawn on 23 August, deployed his two cavalry divisions in two echelons, concentrating both divisions in woods south of the Mezha river, opposite German Ninth Army units. The cavalry penetration focused on the space between two fortified points, Podviaze and Uste. At midnight, the regiments moved forward; their night attack stunned German defenders. With a swift assault, the cavalry group broke through, destroying more than an infantry battalion. Once behind the German lines, the two cavalry divisions were unable to maintain steady communications with each other. This became evident in their poorly coordinated actions which would have been Radzievskii's responsibility as chief of staff. He was learning his trade in combat.

During the raid, 53rd Cavalry Division battled German motorized infantry which were attempting to press the cavalry into a swampy region. German ground attack aircraft twice struck the division. From 23-29 August, the raiding group destroyed German garrisons, headquarters, trains, and storage depots, and disrupted communications between German units. In the area of Guki, Zheliukhovo, they routed the German 430th Infantry Regiment. The Sixth German Army headquarters, discovered by Red Cavalry, quickly fled, spreading panic rumors about a breakthrough of "one hundred thousand Red Cossacks."

Both cavalry divisions traveled nearly 300 kilometers through roadless forest-swampy terrain around Smolensk, in a deep run through the German Ninth Army rear area.

On 29 August, the cavalry group began its return movement. Despite German response with powerful aviation, tanks, motorcyclists with machinegunners and motorized infantry, the cavalry broke through the German front for a second time, returning to friendly territory, 30th Army sector. The raid gave the German Army a healthy respect for Red Army cavalry.

As a result of these activities against the rear areas, the Red Army claimed more than 2500 German soldiers and officers, two regimental headquarters and topographic section of the Sixth Army headquarters, 200 motor vehicles, two tanks, four armored vehicles, four guns, six mortars, and thirty medium machineguns were destroyed. The German command, troubled by the cavalry group raid, halted their advance and withdrew units on the Sixteenth Army front.[1]

The legendary operation of Colonel L. M. Dovator's cavalry group in August 1941 became a premier example for actions against German rear areas. Detailed analysis by the Red Army General Staff War Experience Section used the operation's successes as a model.[2]

In September, Radzievskii, as chief of staff for the 53rd Cavalry Division, organized combat operations characterized by surprise, bold, and highly mobile maneuvers. Often times, Radzievskii's units moved through extremely difficult terrain to achieve surprise. This is the operational environment that honed his combat experience for mobile warfare in the operational depth.

Radzievskii, as the main executor for carrying out the commander's intentions and troop administration, freed the commander to follow the course of combat operations, study the situation, move to critical points on the battlefield, and make decisions. The staff, organized and directed by the chief of staff, prepared assessments and recommended course of actions to the commander. The chief of staff was the only one who had the right to issue orders to troops on behalf of the commander, and by regulation, a chief of staff order issued on behalf of the commander was compulsory for execution.[3]

During the Great Patriotic War, the chief of staff was invested with responsibilities for maintaining high combat readiness of troops and staffs, for organization and support of continuous communications with subordinate and coordinating troops, the next higher staff and adjacent units, for state of reconnaissance, for organization of the commandant's service, for physical security, and for taking measures for all other types of support. Only the chief of staff had the authority to give instructions in the name of the commander to all personnel subordinate to the commander, and, in urgent cases, to make changes on his own in plans made by subordinate commanders. This authority imposed a great responsibility on the chief of staff, most importantly for knowing thoroughly not only the commander's decision, but also his assumptions and his courses of action in the event of possible changes to the situation. The key relationship between commander and chief of staff required an ability to work together and to grasp each other's thoughts quickly.

The chief of staff organized coordination between various army branches and services, and directed their activities towards accomplishing the commander's orders. Chiefs of army branches and services were under obligation to report to the chief of staff on conditions of their troops and supplies. Rear and supply services planned according to chief of staff direction.

In October 1941, 29th and 30th Armies moved back from Belyi to Sychevk; their withdrawal was covered by the 50th and 53rd Cavalry Divisions. The corps had been spread to defend an extensive frontage north of Volokolamsk up to the Volga Reservoir. Cavalry units initially assumed defensive positions, then withdrew themselves to Volokolamsk. Radzievskii had benefited from his combat experience and knew what the cavalry could do, as well as the strengths and weaknesses of the enemy. His priority work was an active reconstitution of units and preparations for future combat.

In mid-November, the Western Front's right wing conducted a counterstrike with General K. K. Rokossovskii's 16th Army north of Volokolamsk. "Despite substantial losses, the 3rd Cavalry Corps was a force to be reckoned with," Rokossovskii observed after the war. "Its officers and men had done a lot of fighting and knew the sound of bullets."[4] Although successful initially from rapid, energetic strikes, the cavalry was thrown back as Germans brought up additional forces. Villages changed hands several times in the fierce fighting.

The cavalry suffered its greatest losses from German air strikes forcing them on the defensive. The action involved Radzievskii's cavalry unit which, located in different areas, was unprepared to complete its assigned missions. The 16th Army actually had insufficient force to complete its attack. The Red Army discovered during these early offensive operations that its mobile forces were too small to fully encircle large German groupings. Consequently, many offensive operations were opportunities lost. It would take the Red Army another two years to build armored forces large enough for major encirclement operations. In the meantime, cavalry units conducted raids similar to Radzievskii's unit, operating on a battlefield increasingly ill-suited for horse units.

In the area of Solnechnogorsk, on 24 November, German forces captured the city and attempted to develop the offensive towards Moscow along the Leningrad highway. The Western Front commander, General of the Army G. K. Zhukov, ordered a counterstrike to recapture Solnechogorsk. Organizing the counterstrike, General Dovator, commander 3rd Cavalry Corps, reinforced his rifle and artillery regiments. The cavalrymen inflicted the counterstrike, enjoying success initially. General I. A. Pliev's cavalry division nearly destroyed the entire 240th Infantry Regiment. The remainder of the cavalry corps advanced forward.

In these audacious battles, Major Radzievskii encountered situations in which he displayed his personal bravery and courage. During the course of fighting, German infantry with tanks penetrated forward cavalry units, advancing towards his division staff location. Radzievskii quickly organized a local defense of the cavalry command post, using corps staff workers and political affairs personnel to occupy fighting positions.[5] Meeting a road block covered by intense, organized fire, German attackers failed to persevere and withdrew, leaving on the battlefield tens of dead machinegunners.

The cavalry corps was unable to fulfill its assigned mission, because of large losses and German superiority in forces. However, it prevented German development of their major attack towards Moscow. The Red cavalry counterstroke bought time for additional reserves to be concentrated on the important axis. Nonetheless, Dovator's cavalry group covered western approaches along the Volokolamsk highway, fighting along side General M. E. Katukov's 1st Guards Tank Brigade. Close coordination between cavalry and tank units was ensured by Radzievskii. The cavalry divisions actions in defensive battles around Volokolamsk in October and November won the divisions "Guards" designation.

In November 1941, Major General L. M. Dovator's 3rd Cavalry Corps became 2nd Guards Cavalry Corps, receiving its guard banner in a ceremony on 8 December. With the formation of the Guards cavalry corps, Radzievskii received promotion to Lieutenant Colonel and became corps chief of staff.

At the end of November, the corps briefly became part of the 16th Army fighting against German forces approaching Moscow from the northwest. After a short respite, the 2nd Guards Cavalry Corps was assigned to reinforce General D. D. Lelyushenko's 5th Army. In early December, the cavalry corps received orders to conduct a strike in the vicinity of Istra with the following mission: 2nd Guards Cavalry Corps with 22nd Tank Brigade and battalion of guards mortars (Red Army name for the famed Katyusha rocket launcher units) attacks through the 329th Rifle Division of 5th Army, from line Apalchschina-Zaovrazhe-Lokotnya move into the penetration and operate in the rear of German forces vicinity of Istra, not allowing the advance of their reserves, destroying withdrawing enemy and preventing their fortifying intermediate defensive lines.[6]

The 2nd Guards Cavalry Corps, now consisting of the 3rd and 4th Guards Cavalry Divisions, 20th Cavalry Division, and 22nd Tank Brigade, concentrated in the area north of Kubinok on 11 December.[7] However, the late arrival and preparation of some units delayed the operation for two days, allowing additional time for offensive preparations. Reinforcing cavalry divisions from Central Asia arrived when the ground had frozen and ice covered the marshes, but the horses had not been reshod for the winter and this hampered the cavalry's movement. In addition, the fresh officers and men lacked the experience of mounted action in broken, marshy and wooded country.[8]

At 0900 hours, Red Army forces unleashed a powerful artillery bombardment. General Dovator, together with the army commander, General D. D. Lelyushenko, were located in the observation post. The generals fol-

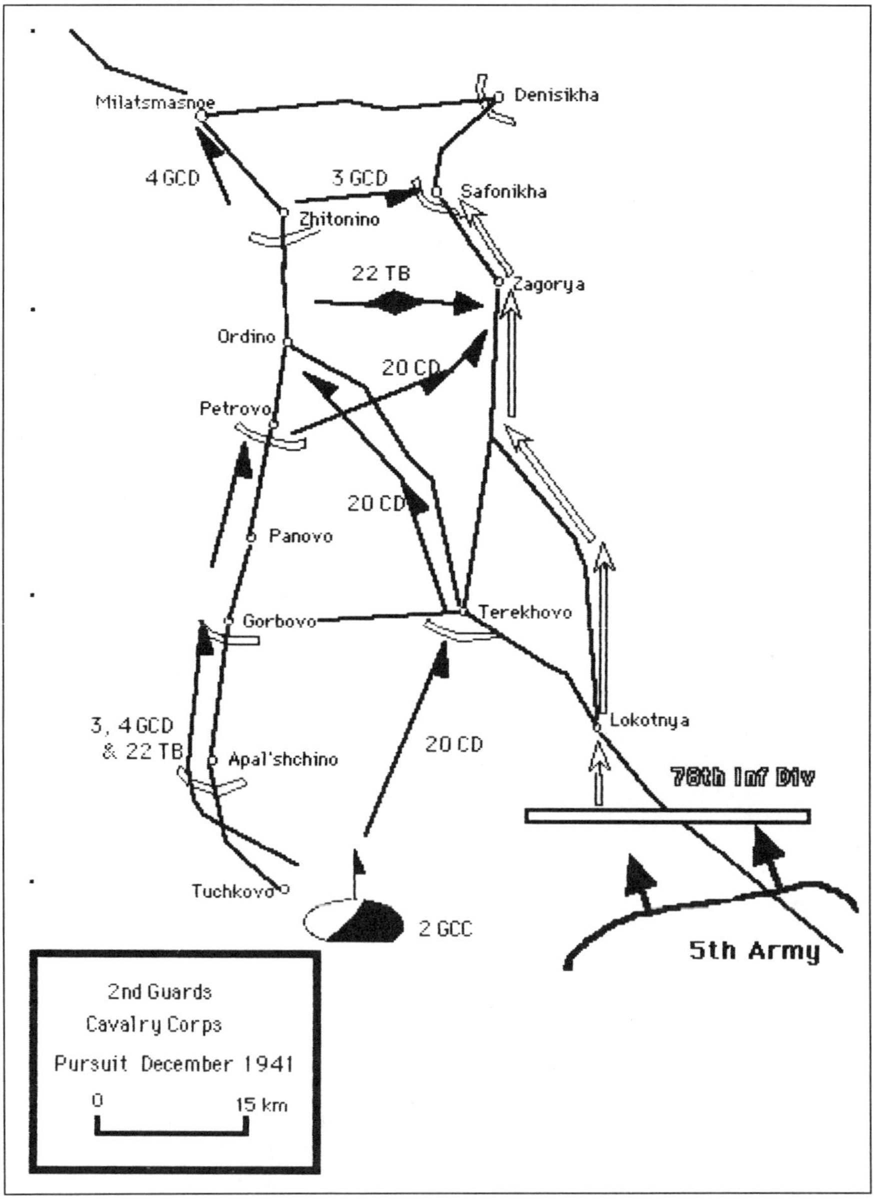

lowed attentively the course of the battle. Snow fell everywhere in drifts; visibility was less than 200-300 meters. Looking through his binoculars and then turning to Dovator, Lelyushenko suggested that it might be better to wait until the defense was fully penetrated. By afternoon, rifle divisions made gains, and, in anticipation of German reserves being brought up, General Lelyushenko decided to commit the cavalry units.

Acknowledging the Army commander's decision, General Dovator over his shoulder passed the order to Radzievskii, "Give the signal for the divisions to move forward!"

Radzievskii passed the signal to subordinate units and watched execution of his staff's plan. Penetrating thick woods, forward cavalry units moving into the rear of the German Fourth Army, attacked its lines of communications and conducted operations in coordination with 5th Rifle Army offensive.

By first day's end, 2nd Guards Cavalry Corps destroyed the German garrison in Apalchschina, advancing in a northern direction, and captured Gorbova. German prisoners indicated that the 78th German Infantry Division was withdrawing from the area of Lokotnia to Terekhovo, Zagore, Safonikha. General Dovator decided not to let enemy units withdraw westward. As German units attempted for two days to slip westward, the 20th Cavalry Division pursued relentlessly the tails of retreating columns, capturing material and beating enemy rear guards. At the same time, the corps' left flank forces, consisting of the 3rd and 4th Guards Cavalry Divisions and 22nd Tank Brigade, moved along wooded parallel roads to the German's withdrawal path, destroying their garrisons in villages and towns along the road. These actions stalled 78th infantry Division units, repelling their attempts to trickle to the west.

On 16 December, the 78th Infantry Division, moving in a column from Zagore to Safonikha, the length of which was greater than 6 kilometers, was encircled and engaged decisively by 2nd Guards Cavalry Corps. General Dovator threw the 3rd Guards Cavalry Division and 22nd Tank Brigade from the west and northwest against the trapped Germans. While 20th Cavalry Division struck the column's tail, 4th Guards Cavalry Division remained as a screen on the northwest direction in case the enemy succeeded in filtering through north of Safonikha.

The German column attacked simultaneously by cavalry and tanks split into smaller defensive segments, with the greater portion destroyed. Approximately a battalion and half of German infantry took up positions in Safonikha and Denisikha. In a desperate defense, German remnants lasted until the morning of 17 December. The 78th Infantry Division was utterly routed; only a few soldiers ran off and hid in nearby forests. Guards cavalrymen captured 82 guns, 23 mortars, 143 machineguns, 23 tanks, 2 armored cars, 431 motor vehicles, 89 motorcycles and a large quantity of weapons, ammunition, and other materiel.[9] With flank attacks, Red Cavalry units inflicted severe losses on the division. Soviet sources claiming the German infantry division lost 70% of its personnel and created favorable conditions for the advance of General Lelyushenko's 5th Army.[10]

After the 78th Infantry Division, the cavalry encountered the 252nd Infantry Division withdrawing along the Istra-Ruza road. Again, the horse force inflicted heavy losses in combat equipment and guns on the beleaguered infantry unit.

"The deep snow and severe frosts prevented us from engaging in enveloping movements away from the roads to cut off the enemy's routes of withdrawal, and German Generals, instead of blaming the Russian winter for their defeat, should perhaps be grateful for the harsh weather that enabled them to retreat with less losses than they would otherwise have suffered," Rokossovskii wrote in summing up the operation.[11]

On 19 December 1941, 2nd Guards Cavalry Corps advanced along the Volokolamsk-Mozhaisk highway. A liaison officer from corps headquarters located General I. A. Pliev, commander 3rd Guards Cavalry Division, and informed him that General Dovator had been killed. Lieutenant Colonel Radzievskii asked for his presence at the command post.

In the first week of January 1942, the Western Front prepared a new operation with the objective of destroying the Sychev-Rzhev German grouping. In 20th Army sector, 2nd Guards Cavalry Corps reinforced with 22nd Tank Brigade and five ski battalions for the echelon for developing the breakthrough.[12] After a 35 kilometer night march, the corps concentrated in the area of Matrenino, Ivantsev, in preparation for commitment to the penetration. The plan was to commit the reinforced cavalry corps on the second day of the operation.

On the morning of 10 January, a heavy snow fell. Visibility was poor. Air support could not operate, and the artillery preparation was weak. Front forces jumped off in the attack at 0930 hours. Radzievskii remembered, "Storming from bunkers, the division's battalions advanced slowly forward. From insufficient ammunition the artillery weakly supported attacking infantry and tanks."[13] The advance was only 2-3 kilometers at first. For the next day and a half, Front forces overtook only another three kilometers of German defense. Deep snow made advance difficult. The Army commander decided to commit the cavalry corps. However, General Zhukov, the Front commander, prohibited commitment of the primarily cavalry mobile group until penetration of the tactical defensive zone. The 20th Army by evening of the 12th was at a depth of 6-7 kilometers.

Three mornings after the attack began, Pliev's cavalry corps was committed in a penetration at Shakhovskuyu. Two days later, Zhukov directed Pliev's cavalry towards Seredy and Gzhatsk to capture German units which were beginning to withdraw. General L. M. Sandalov, chief of staff of the 20th Army, drove with orders to the cavalry corps command post. There, he found Radzievskii and repeated Zhukov's orders. Pliev, who returned to the command post during Sandalov's visit, quickly made the decision to turn the corps.[14] By 14 January, Pliev's sabers seized Seredy driving towards Gzhatsk, but German resistance stopped the cavalry corps. "The tempo of our offensive did not exceed the tempo of the enemy's withdrawal . . . The mobile group did not possess sufficient strike force, since it did not have the necessary number of tanks and great difficulty of maneuvering in very deep snow," Radzievskii noted.[15] Needing rest, the cavalry was placed in reserve.

The 20th Army commander, in the period from 13 January through 15 February 1942, directed the 2nd Guards Cavalry Corps (3rd and 4th Guards Cavalry Divisions, 20th Cavalry Division and 22nd Tank Brigade), which was under his operational subordination, to attack the villages of Chukhalovo, Bol'shie Tresaly, Bykovo, and others, which had been strongly fortified by the Germans. Since the 3rd and 4th Guards Cavalry Divisions possessed no division artillery and very few mortars, the corps was forced for more than a month to conduct stubborn battles behind a fortified line, and suffered enormous losses without achieving tactical success.[16]

In the winter months of 1941-42, the Red Army held Moscow out of reach from advancing German pincers and conducted successfully, but limited, counteroffensives. By March 1942, the Kalinin and Western Front forces advanced to the line of Velikie Luki, Vyelizh, Belyi, Rhzev, Gzhatsk, Spas-Demensk.

Lieutenant Colonel Radzievskii learned from experience in offensive operations in the winter and spring of 1942 that cavalry units found local resources an unreliable means for sustaining themselves, considering the system of pillaging and destruction practiced by the German Army. In January-February 1942, the 2nd Guards Cavalry Corps on the 20th Army front conducted a drawn out battle for fifteen-twenty-five days in dismounted combat formations against German infantry, who had occupied fortified villages. As chief of staff, Radzievskii dealt with coordinating these operational constraints. The horses remained in villages partly destroyed by the withdrawing Germans, where, because of the uninterrupted German Luftwaffe activity, the water supply was irregular and forage was supplied with large interruptions. As a result, a large withdrawal of horses took place and there was nothing with which to replace them.[17] Red cavalry, forced to forage in great quantities, found itself reduced significantly in their mobility and combat capability.

The ruggedness of Red cavalry horses was impressive. The horse, a shaggy small Kirkhil equine from Siberia, could withstand temperatures of 30 degrees below zero. They could negotiate dense forests and muddy roads that tanks often could not traverse. Despite their operational independence, the cavalry division required special handling and support.

"It is necessary to periodically give cavalry a 'breathing space' to bring the horses to order and reestablish forces before a new operation," a Red Army General Staff Study concluded. "As a rule, in conducting battle in dismounted formation, the horses remain in the cavalry's immediate rear with a small number of horse holders. Here care and maintenance of the horses is reduced and the cavalry is quickly reduced in its combat composition, since ill horses are sent to the division and corps rears with a certain number of men."[18]

Red Army staff analysis determined that the duration of the 'breathing space' was solely dependent upon the situation at the front, the condition of the cavalry after the conducted operation, the degree of intensity of forthcoming activities, and similar stresses of operations. In any case, time had to be given to the cavalry so that men and horses who had gathered during battles could be brought up from the rear, personnel and material could be obtained and distributed, and horsed could be reshoed.[19]

In the spring and early summer 1942, the Western and Kalinin Fronts assumed their former line, organized and improved their defenses. German units took actions which Red Army leadership believed indicated preparations for resuming their offensive for Moscow. Throughout the summer, with preparations for defense and the arrival of new equipment and personnel, Fronts reinforced and fortified positions on a direct axis to Moscow.

Aleksei Ivanovich's first combat experience began with the 53rd Cavalry Division raids into the German rear area around Smolensk. Then, as chief of staff for the 2nd Guards Cavalry Corps, he participated in defensive battle and 150 kilometer raid around Moscow.

From early summer 1942, Radzievskii transferred as chief of staff to the 1st Guards Cavalry Corps. He joined the now famous cavalry corps after it had returned from its operations behind the German lines in the area of Vyazma after winter campaigns. The operation took its toll on subordinate units, requiring several months to rebuild battered cavalry formations.

After resumption of a German offensive in May and June 1942, fast-paced panzer and panzergrenadier advances left cavalry units with difficulty in regrouping and maneuvers to keep pace with the battle. Cavalry corps actions were minimal. The 1st Guards Cavalry Corps was held in reserve to guard against further attempts by German forces to capture Moscow. Summer and fall fighting saw the 1st Guards Cavalry Corps committed to combat operations in the Kaluga area, where it fought defensive battle stopping German advances and stabilizing this portion of the German-Soviet front.

By 1943, with stabilized fronts and Red Army attempts to resume their offensive, the cavalry unit role in the development of offensive actions and the number of tank and mechanized units increased. The cavalry, again, began to operate in German unit rear areas.

Wanting to maintain offensive momentum after victory at Stalingrad, the Stavka decided to expand and continue its winter offensive. While Southwestern and Southern Fronts planned an operation through the Donbas toward the Dnepr river, the Voronezh Front received instructions for an advance on Kursk and Kharkov.

The Voronezh Front, consisting of five armies, received the 1st Guards Cavalry Corps as a reserve. The 38th and 60th Armies had the mission to attack towards Kursk; the 40th, 69th Rifle, and 3rd Tank Armies, toward Kharkov. Their offensive began on 2 February 1943. Initial successes by the Red Army blinded Soviet leadership to ominous positionings by German SS panzer units on the flanks. German commanders were intent on a counter-offensive. Red Army units were unable to continue an easy attack.

On 5 February, General Rybalko's 3rd Tank Army fell behind the Front's planned schedule of advance, and he sought ways to overcome the Northern Donets River. Sweeping further to the south, the tank army continued its advance after a four-day battle with SS Das Reich Division, detracting from the army's main mission to cross the Donets and seize Kharkov.

The following day, Red Army forces seized Balaklei, Barvenkov, and, for developing the offensive, 1st Guards Cavalry Corps, under the command of Major General V. K. Baranov, moved along the railroad from Sukhinichei. Front successes against the SS Panzer Corps forced them back to the Kharkov outskirts by 13 February. Strong advances by Red Army forces threatened many of the German units' rearward lines of communications. The struggle at the operational level gave a transitory opportunity to the Germans. While

Red Army units made vast territorial gains, they failed to trap and destroy major German units. Pulling back German units to shorten frontage in southern Russia, Army Group Don created spare forces for use in a counteroffensive. Additionally, by ordering seven fresh divisions transferred from the West to Army Group South by the end of the first week in March, Hitler provided the Army Group Commander, Field Marshal von Manstein, with the means to stem the Red Army advance.

On 19 February, Army Group South's counteroffensive began. In a series of swift blows, the army group eliminated Red Army threats to crossings on the Dnepr river. The Southwestern Front advance to the south of the Voronezh Front had been shattered in the first days of the German counteroffensive. By 25 February, von Manstein issued orders that prepared German forces to seek destruction of the Southwestern Front's advance to Kharkhov. Attacking against the southern flank of the Southwestern Front, the German's Fourth Panzer Army made significant advances, and in five days cut through elements of the 1st Guards Army and destroyed units of the 3rd Tank Army. The German advance retook Kharkov.

By 28 February, the scope of the Red Army disaster became apparent to General Vatutin, Voronezh Front commander, and the Soviet High Command. The 3rd Tank Army was assigned to the Southwestern Front and to attack southward to stop the German advance. By 5 March, 3rd Tank Army, worn down by the constant fighting, was chewed up by the SS Panzer Corps. The destruction of the 3rd Tank Army south of Kharkov, and the relentless German advance in mid-March grew rapidly into a Red Army catastrophe.

A major effort in reordering and regrouping its forces in the Voronezh and Southwestern Fronts was necessary. On 13 March, the situation in Voronezh sector took a serious turn for the worse. A German panzer strike from the north sealed 3rd Tank Army's fate. General Vatutin reinforced the 3rd Tank Army, on 13 March, with the bulk of 1st Guards Cavalry Corps, with two understrength cavalry divisions (7th and 2nd). The cavalry corps manned defensive positions along the Rogan river, on 13 March, and began delaying operations with 113th Rifle Division against elements of SS Totenkopf and Das Reich Divisions. After three days of fierce fighting, Totenkopf pushed 1st Guards Cavalry Corps' two divisions out of Chuguyev back to new defensive positions west of the Northern Donets River where the cavalry corps' left flank was joined by another cavalry division. By evening 14-15 March, the tank army received permission to abandon Kharkov, an order too late.

A similar disaster befell 69th and 40th Armies. The 69th Army commander maneuvered his divisions to a new defensive position north of Kharkov, but the German's Grossdeutschland Division split the armies from each other. The 69th Army's left flank was covered by the 1st Guards Cavalry Corps, which the Stavka had just committed east of Kharkov to cover the 3rd Tank Army withdrawal from the city.

Two days later, Fourth Panzer Army crushed the last resistance in Kharkov, remnants of Red Army forces on the afternoon of 16 March after heavy fighting withdrew northeast, rejoining Soviet main forces through the 1st Guards Cavalry Corps.

The SS panzer divisions, while fighting 1st Guards Cavalry Corps, shifted northward to new assembly areas in order to prepare for an advance on Belgorod the following day. The XXXXVIII Panzer Corps's 6th and 11th Panzer Divisions assumed control of the sector, and, over the next four days, slowly drove the 1st Guards Cavalry Corps in very heavy fighting against the Northern Donets river. On 26 March, 1st Guards Cavalry Corps abandoned its last positions west of the river. Red Army command with exhausted troops began hastily building defenses on the Donet's east bank to hold their gains.

Destruction of the 3rd Tank Army and absence of viable Front reserves collapsed the Voronezh Front's southern flank, opening of a gap between the 69th Army and tank army remnants. In a rare instance of cavalry forces engaging armored units, 1st Guards Cavalry Corps delayed against German forces for three days, from east of Kharkov back to the Northern Donets River. Radzievskii experienced first hand the results of over extended operations and the vulnerability of forces unprepared for German panzer counterstrikes. The operation held valuable lessons for him.

The 1st Guards Cavalry Corps, in July 1943, stayed in the Voronezh Front which became the 1st Ukrainian Front in October 1943. At the same time, the rapidly rising Radzievskii received promotion to Major General. Red Cavalry had contributed significantly in rolling back the German invaders. Manstein wrote: "A Soviet cavalry division can move, in its entirety, a hundred kilometers in a night - and that at a tangent to the axis of communication."[20]

In November 1943, Radzievskii participated in his last major operations as a cavalry corps chief of staff in the 1st Ukrainian Front's Kiev offensive operation. Committed from the Lyutezh bridgehead north of Kiev on 6 November, 1st Guards Cavalry Corps attacked southwest to cover the 3rd Guards Tank Army flank, as it began its advance into the operational depth.

The 1st Ukrainian Front commander, General Nikolai Vatutin, committed 1st Guards Cavalry Corps into a gap between the German 13th and 7th

Army Corps. The cavalry corps advanced directly on Zhitomir. By 12 November, the 1st Guards Cavalry Corps supported by a full rifle corps of the 38th Army entered Zhitomir, but in doing so it became a lucrative target for German counterattacks from the south and southwest.

On 16 November, German attacks continued moving northward towards Brusilov and moving towards Zhitomir. As 7th Panzer Division and XXXXVIII Panzer Corps moved northwest, they threatened to cut off cavalry elements in the city. The following day, 7th Panzer Division counterattacked 1st Guards Cavalry Corps in the city.

The next evening, Red Army forces within Zhitomir found themselves encircled by 7th Panzer Division from the south and 8th Panzer Division from the north. However, the greater situation posed a problem for the Germans. It is one thing to encircle with armor, but it takes an infantry encirclement to hold tightly trapped forces. Because of the paucity of German infantry units, most of these encirclements were porous. Many Red Army units were able to get out, although, often minus their equipment.

By 19 November, German forces had cleared Zhitomir. Cavalry divisions in Radzievskii's corps fended for themselves, slipping northeast into 60th Army, which sent its rifle reserves to stabilize the situation. The 1st Guards Cavalry Corps collected its units, reestablishing command and control while a 60th Army rifle corps assumed the fight. The deteriorating situation required cavalry corps remnants back into the line to stem German advances within a week of their escape from encirclement.

On 6 December, German panzer units brought into the area nearly encircled, again, 1st Guards Cavalry Corps northeast of Chernyakov. In a fighting withdrawal eastward, the cavalry corps held the line against powerful 1st and 7th Panzer Division thrusts. By mid-month, the cavalry corps found itself still squared off against the 7th Panzer Division. On Christmas Day, the 1st Guards Cavalry Corps pulled out of the line and moved north. The corps received no respite throughout January 1944, participating in 1st Ukrainian Front operations.

In battles 1942-43 around Kharkov, Northern Donets, forcing the Dnepr river, liberating Kiev, Zhitomir, and other Ukrainian towns and villages, Radzievskii operated as the chief of staff for the 1st Guards Cavalry Corps. These latter operations gave him great experience in the preparation and coordination for the execution of breakthroughs in German defenses and development of the offensive in the operational depth far from main forces.

On 22 February 1944, General Major Radzievskii was designated Chief of Staff for the 2nd Tank Army. The chief of staff plays the role of work organizer for unit commanders and staffs. Smooth, coordinated functions of all commanders and staffs, especially the staffs, depends on his training, experience, and skill in directing the work of his subordinates. His personal qualities for organizational management, capacity to find in each situation the link in the chain that will decide success of the matter, and to mobilize all forces to achieve the main objective, characterize the chief of staff as the team leader.

In his war memoirs, Marshal of the Soviet Union S. S. Biryuzov writes the following about the chief of staff: "The commander must believe in his chief of staff as he would in himself. It is impossible to work otherwise. The chief of staff is not simply an executive agent. He is one of the commander's closest assistants and must have a creative mind and disposition. The chief of staff, on the basis of the general concepts of the commander, considers all the details of the situation with his staff and prepares motivated proposals. It is through him that the execution of orders is monitored and command and control is provided."[21]

The chief of staff works directly with the commander on making the decision, he calculates the time allocated to organization of combat operations, and he develops those very important operations documents that usually require coordination among various control organs.

Beginning in 1942, mandatory organization of troop coordination directly on the ground was a constant requirement. The whole concept of the organization of the staff and its function was well defined in the Red Army Field Service Regulation 1942.

"The Chief of Staff is the main agent for carrying out the commander's will and troop administration. In the commander's absence, he becomes his deputy. He organizes and directs the entire activities of headquarters and is fully aware of all the intentions and plans of the commander.

"The Chief of Staff's duty is to provide the commander with full opportunity to follow the uninterrupted course of combat activities of the troops; to study the situation and to make decisions. The Chief of Staff must organize his work in such a way as to be able to direct all efforts towards the solution of basic problems of troops direction. The chief of staff should always know the situation and be prepared to give his appraisal and suggestions at the commander's request.

"The chief of staff is the only one who has the right to issue orders (or instructions) to troops on behalf of the commander. An order (or instruction) issued by the chief of staff on behalf of the commander is compulsory for execution.

"The Chief of Staff, charged with organizing the conduct of and the cooperation between army branches, supervises the work of heads of army branches and services, and directs their activities towards carrying out the commander's orders.

"Heads of army branches and services are under obligation: to report to the chief of staff on the condition of their troops and supplies; to work out under the direction of the chief of staff all measures pertaining to their branches and services; and to ensure the execution of the commander's orders.

"The chief of staff, in planning the organization of the rear, directs the activities of heads of supply services and instructs them with regards to such organization and servicing. Heads of supply services and assistants to the commander of the unit's supply and technical services lay down their plans for supplies and repairs according to the directions of the chief of staff. They must report at all times to the latter on matters pertaining to the servicing of the rear."[22]

Upon arrival at the 2nd Tank Army headquarters, Radzievskii began immediately to plan combat actions for the Uman-Botoshansk offensive operation. In a short time, he knew the combat situation, understood the unit's mission, and learned thoroughly the enemy. He required from subordinates an exactness and objectivity in staff actions, constantly reminding them that in war, small errors or misrepresentations exact a severe price in soldiers' blood.

Radzievskii emphasized development of the artillery plan and considerations, if weather permitted, for the use of aviation in the main sector attacks. He organized an artillery density of 30-35 guns and mortars per kilometer of front which ensured a good chance of success in tank army actions for capturing on the move hastily occupied enemy defenses.[23]

In preparation for an operation, a key measure by the tank army staff was coordination with Front aviation for the entire depth of the operation. The number of aircraft allocated for tank army support increased as the war years grew, and, by the end of the war, averaged up to one fighter and ground attack air corps per army. Operational aviation groups went to the headquarters of an army and corps for coordinating issues. All details of joint operations for the time of the operation were refined and coordinating signals were specified - calls for aircraft, mutual identification, target designation on the battlefield.[24]

The tank army assumed the breakthrough in the 27th Army sector, which had carried out only 40% of its mine clearing operation. Despite slow clearance through the tactical zone, the 2nd Tank Army's tank corps forced the Gornyi Tikich river and fought for Uman. Developing the offensive, they crossed over the Southern Bug and Dnestr rivers, and struck towards the town of Yassy.

The chief of staff, to ensure a cohesive command and control, knows missions assigned to corps subordinate units, knows the current situation, status, and capabilities of friendly and enemy units, and quantitative and qualitative correlation of forces. At any point in the operation, he is ready to draw conclusions from an assessment of the situation and make recommendations in alternative courses of action in supporting combat operations and organizing command and control.

In the course of the offensive, Radzievskii used any occasion to drive to tactical units. Forward, he could see and feel the general character of the fighting.[25] He taught subordinates how operationally to choose in the flow of communications and to report that which was necessary and valuable. As a result, his staff prepared exact, well-grounded, and timely information, allowing correct decisions to be made.

By regulation, "headquarters officers must set an example to the rest of the headquarters personnel; they must be familiar to a high degree with the character and nature of modern warfare, and be able to conduct staff work intelligently and expeditiously."[26]

During the war years, specific work carried out by the staff demanded from officers profound analytical thought, industriousness, patience, and the ability to spot a development trend in one or another event and foresee what this development could lead. All of this required special attention in manning staffs and a consideration of not only the professional qualities of officers but also personal character traits and their inclination. While individual staff personalities from the 2nd Tank Army have remained anonymous, it is worth noting the approach to allocating work in the staff proceeding from the psychological features of the individuals as described by the former chief of staff of the 8th Guards Army, Major General V. A. Belyavskii: ". . . LTC S. I. Martsenko, a calm and cool-headed officer, was basically concerned with painstaking work of compiling calculations and plans tables. The operator, Major K. F. Petrov, a person of great erudition and a sharp mind, was irreplaceable on questions requiring an immediate response. We were repeatedly aided by the pedantic efficiency and sound work . . . of Major P. F. Zheltov. When it was a question of drawing up documents, first violin was played by the master of clear and terse formulations and the superior graphic artist, Major I. A. Ivashchenko." As a result, the persons who differed in terms of the type of character complimented one another and comprised that close-knit collective without which effective work by all elements is inconceivable.[27]

A radio bureau receiver for each tank army was allocated for communication with the General Staff Headquarters. The presence of dependable communication maintained with the Red Army General Staff Headquarters made it possible to use it at the needed moment as an alternate link for communication with the Front staff. In the Uman-Botoshansk operation, when 2nd Tank Army's communication with the front staff was broken, it was maintained by way of the radio center of the Red Army General headquarters.[28]

In the 2nd Ukrainian Front's Uman-Botoshansk operation from March to early April 1944, the plan concept had a powerful strike in the general direction of Uman, destroying the Uman-Khristinovka German grouping and reaching the Dnestr River. Between 5 and 7 March, 2nd, 5th, and 6th Tank Armies with infantry, artillery, and aviation support, penetrated the German tactical defense zone. The armored forces crossed the Gorny Tikich River, and pursued retreating German forces.

In the course of 2nd Tank Army advance to the Gorny Tikich, reconnaissance located intact bridges near Buka and Berezovka. Based on reconnaissance reports, forward detachments from 3rd and 16th Tank Corps were dispatched to seize the bridges on the move and prevent enemy destruction of them. Accompanied by an assault force of combat engineers and motorized riflemen, the forward detachments reached the river in the dark early hours of 7 March, behind withdrawing German units. The detachments conducted a hasty crossing, captured the bridges, and dug-in small bridgeheads on the west bank. The bridge captured by the 16th Tank Corps forward detachment in the vicinity of Buka had suffered little damage, while the bridge which was captured by the 3rd Tank Corps forward detachment near Berezovka was severely damaged. The bridge near Buka was quickly restored by the 205th Combat Engineer Battalion and 15th Motorized Rifle Brigade. By 7 March, the entire corps was able to cross the river over it. Responding to an order from the army commander, 3rd Tank Corps maneuvered along the river from the Berezovka area, crossed the river on the bridge restored by its neighbor on the right, and resumed pursuit of German forces in its previously designated sector.[29]

On 10 March, 2nd and 5th Tank Armies liberated Uman, capturing over 500 serviceable tanks and more than 350 guns. With the fall of Uman, German forces began withdrawing to the Southern Bug River.

Pursuing German troops, on 11 March, the tank army reached the Southern Bug and carried out a hasty crossing, compelling the German command to withdraw to the Dnestr River.

Throughout the war, bridging resources for tank armies and their corps could not independently support tanks crossing of mid-sized, much less large, water obstacles. The 2nd Tank Army, during its forcing of the Dnestr River in the Uman-Botoshansk operation, had tanks wait eight days for a bridge 267 meters long to be built. As tank armies advanced westward into Europe, bridging rivers became a significant problem. Radzievskii's staff increasingly coordinated bridging assets for continued advance to avoid delays in the operational depth.

In the summer of 1944, the Red Army sought to destroy German Army Group Center with the Belorussian offensive operation. The 2nd Tank Army was part of the 1st Belorussian Front on the southern, or left wing, of this strategic operation. The general concept of the operation was a strike by the 1st Belorussian Front around the Brest fortified area from the north and south to destroy German forces at Lublin and Brest, and to develop the offensive to the west toward the Vistula River.

In accordance with this concept, 2nd Tank Army commander, chief of staff and staff worked planning the tank corps attacks, coordinating their actions with rifle armies in the first echelon and with cavalry corps that would operate on their flanks.

The commander's methods in decision-making and assigning missions and the army staff's methods in collection and analysis of information for decision-making and planning depended on a considerable number of factors, such as time available, condition of the army, nature of the combat missions, operational training and the army's combat experience of the army.[30]

"The commander should make his decision in the presence of the chief of staff and chief of operations. The chief of operations copies the commander's decision verbatim," regulations required.[31] If the commander makes a decision outside of the headquarters, it is transmitted by the accompanying staff officer to the chief of staff for immediate communication to the troops. The commander himself issues the order to the command of those formations in whose area he happens to be.

Once a decision is made, the chief of staff advises the chief of operations branch of the following: in what sequences, to whom and by what means the preliminary instructions are to be transmitted; the time limit for preparing the field order and in what sequence it is to be transmitted to the troops.

In planning, Radzievskii's touch was evident in attempts to achieve the ultimate in penetrations, a "chistya" (clean) breakthrough. He wrote after the war, "Our basic attention was given to securing the tank army's commitment to the fight. In order not to suffer unnecessary losses in tanks, we organized an aggressive and reliable artillery cover. The method was a series of artillery concentrations to a depth of 15 kilometers. Then, across each 1.5-2 kilometers a designated fire line was established for a total of eight, on which, for all potential strongpoints and enemy centers of resistance, were marked sectors of concentration fires. Availability of the fire scheme to artillerists and tankers permitted easily calls for artillery fire along any sector."[32]

The tank army command post, headed by the chief of staff, organized the advance of troops to the line of commitment and subsequently relocated with the second echelon or reserve at the commander's direction. As the command post column approached the line of commitment, the commander left the observation post and controlled subsequent combat operations from the main command post.[33] Commanders had to be accessible to senior commanders.

The commander and chief of staff should be located together, if possible, at the center of the headquarters. The operations and intelligence sections, as well as the duty officer, should be placed near the commander and chief of staff.

War experience showed that the tank army command post, advancing at an average rate of 30-40 kilometers, relocated once a day, and tank corps command posts one-two times, with the distance of the relocation at single bound reaching 30 kilometers and more.[34]

Army command post communications with the second echelon commanders used wire telecommunications (mainly telegraph), radio, and message delivery. For radio communications, the army's signals officer allocated radio sets for elements of rear service, and installations used mainly telephone lines. Near the army rear command post, landing strips were equipped for liaison aircraft, and rear area message centers collated.[35] To save time, a common practice dropped message pouches from U-2, bi-planes at the location of tank and mechanized corps and tank armies headquarters and, sometimes, at brigade headquarters.[36]

Diagram orders and map orders were used for tactical control and passed via liaison officers. Brigade and corps headquarters were authorized to send reports in the form of diagram-reports or reports on maps. Graphical reporting reduced document drafting time, and, consequently, speeded up orders delivery to subordinates and reports to higher headquarters.[37]

On 11 July 1944, the offensive began with a significant advance by first echelon rifle armies of the 1st Belorussian Front. By the third day, they had advanced nearly 70 kilometers, creating a wide gap in the German line. Cavalry and tank corps captured three bridgeheads over the Western Bug River. By 20 July, the 2nd Tank Army began moving towards the Western Bug, and two mornings later, it received the mission to commit through 8th Guards Army sector and seize Lublin by 23 July.

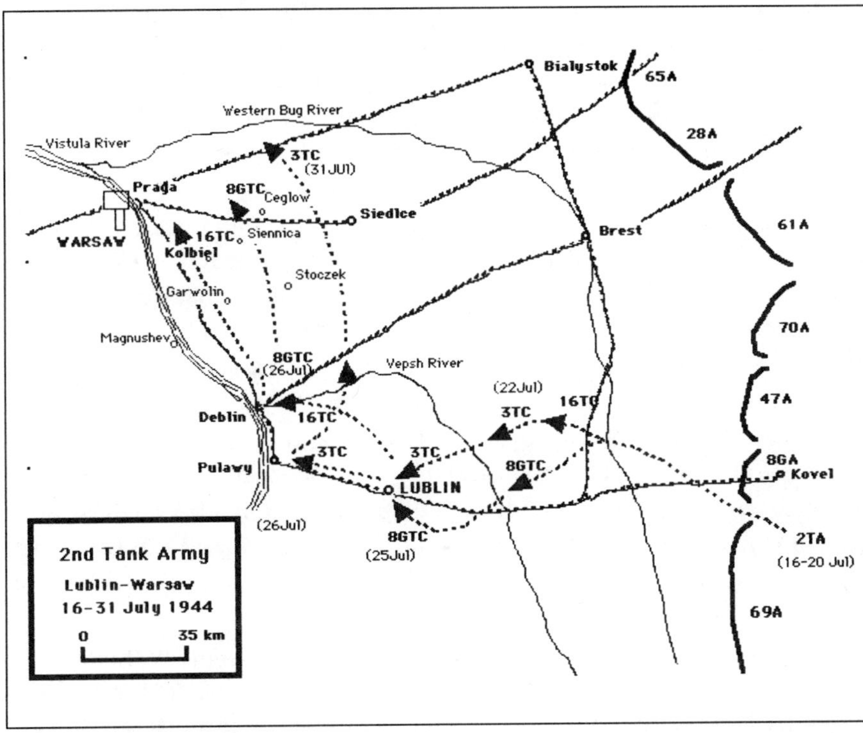

At night on 21 July, the tank army's corps crossed into the bridgehead over the Western Bug. General Semen Bogdanov's concept for his tank army was at dawn to pass 8th Guard Army forces and advance aggressively towards the Vepsh River, forcing it from the march. Going around Lublin from the north, west and east, the tank army in coordination with the 7th Guards Cavalry Corps struck simultaneously on several axes to seize Lublin.

On 22 July, in the Army command post, Radzievskii received the first member of the Front's Military Soviet, N. A. Bulganin, who announced that Stalin asked to pass that Lublin be taken by 25 July. Radzievskii reported that Lublin would be captured not later than 23 July. "Our promise we fulfilled. At night on 23/24 July, Lublin was taken," Radzievskii wrote.[38]

Normally, tank armies bypassed centers of resistance in the operational depth, depending on the nature of their tasks and German strength for resistance.[39] However, war experiences revealed that it was not always advisable to avoid major cities. Some cities turned out to be main objectives for tank army's actions, i.e., the task given 2nd Tank Army to seize the city of Lublin.

General Bogdanov, wounded in the fighting on the western edge of Lublin, was evacuated to Moscow for hospitalization. Radzievskii, at the age of 33, assumed command of the tank army.

With the fall of Lublin, the way was open towards the Vistula River. German forces managed to destroy crossings over the Vistula in the vicinity of Deblin and Pulawy. "Unfortunately," Radzievskii recalled, "we did not have the crossing means and could not manage to force the river from the march because of strong Germans fire."[40]

Radzievskii was ordered to resume the offensive with his main forces in a northerly direction on the morning of 27 July with the mission to capture the Stoczek-Garwolin line by the end of the day. Receiving special instructions for one tank corps, he defended on the right bank of the Vistula from Rytsitsa to Deblin. After securing the its flank, the army advanced in the direction of Warsaw, capturing Praga (a suburb of Warsaw) and cutting the Byalistok-Warsaw and Siedlce-Warsaw rail trunklines.

In order to protect the exposed flank of his Brest grouping, commander Army Group Center moved 5th SS Viking Panzer Division from Brest to the vicinity of Siedlce and the Hermann Goering Panzer Division, which had just arrived from Italy, with the 73rd Infantry Division from Warsaw to the Stoczek-Garwolin line.

Radzievskii passed the following missions to his troops in accordance with the adopted plan: the 3rd Tank Corps was to breach German defenses on the Stoczek-Kamenka line, destroying opposing enemy units and capture the Lyatovich-Gozd line and, subsequently, develop the offensive in the direction of Lyatovich and Ceglow; 8th Guards Tank Corps was to breach German defense on the Kamenka-Garolin line on the move, destroy the opposing enemy and capture the Gozd-Puzhnuvka line, and then exploit the offensive in the direction of Lyubitse.[41]

The 3rd Tank Corps advanced from its assembly area in the direction of Stoczek on three routes. Using a large number of reconnaissance units in the Stoczek sector, the corps commander discovered, while far from German units, their defensive system. Upon arrival, the corps commander decided to penetrate German defenses in the Stoczek sector with an attack from the march. He ordered 50th Tank Brigade and 1,107th Self-Propelled Artillery

Regiment to continue advancing on Stoczek from the south, while the corps' main forces struck in the direction of Lopavyanka (west of Stoczek). Breaching German defenses in this area on the move, the corps destroyed defending German units around Stoczek. The attack on Germans in the vicinity of Stoczek by 3rd Tank Corps units was supported by 6th Air Army bomber strikes. German units began withdrawing to the north and northwest. With the beginning of the German withdrawal, forward units of the 3rd Tank Corps pursued them in the direction of Ceglow, where the forward detachment sent out by the 51st Tank Brigade carried out a hasty crossing. The detachment held until the approach of the corps' main forces.

The 8th Guards Tank Corps advanced on two routes to the left of the 3rd Tank Corps. German units offered stubborn resistance as 28th Guards Motorized Infantry Brigade approached crossings over the Vilga River south of Garwolin. In the afternoon, 6th Air Shock Corps pounced on approaching German reserves in the vicinity of Minsk-Mazowiecke, and garrison defenders in the vicinity of Garwolin. Capitalizing on air strike results, the corps main forces, 59th Guards Tank Brigade and 301st Guards Self-Propelled Artillery Regiment, by-passing Garwolin on the east, carried out a hasty crossing of Vilga River and captured Gloskow (east of Garwolin). Following the capture of Gloskow, the corps turned westward and cut the Garwolin-Warsaw highway, moving from the northwest to the rear of German defenders in the vicinity of Garwolin. In order to exploit the offensive in the direction of Parysow (northeast of Garwolin), commander, 8th Guard Tank Corps, committed his second echelon, 58th Guards Tank Brigade and 1,817th Guards Self-Propelled Artillery Regiment, to the breach east of Garwolin. These units advanced successfully in a northerly direction. The 28th Guards Motorized Rifle Brigade, which was advancing on Garwolin for the south, forced the Vilga river and broke into the city.

In one day's advance toward Warsaw, 2nd Tank Army destroyed the 73rd Infantry Division and Hermann Goering Tank Division covering forces on the Stoczek-Garwolin line. They moved forward 50-55 kilometers, completing its day's mission. Simultaneous penetration of German defenses in several sectors paralyzed the entire system of enemy defenses, broke-up the enemy forces into separate, isolated groups, and promoted their swift destruction in detail. This method of penetration made it hard for German commanders to determine the axis on which the army was to make its main strike, preventing them from offering organized, coordinated opposition in threatened sectors. In addition, the swift strike made it possible to avoid high concentration of troops in the army's zone of advance, and reduced the relative damage suffered by units and formations from German air strikes.

The 2nd Tank Army smashed its way into the enemy's rear area to a depth of 125 kilometers, reaching approaches to Praga. This significantly assisted troops of the Front's right wing and center in destroying the German defenders at Bialystok and Brest.

Beginning 27 July, 2nd Tank Army moved north. Reconnaissance reports on German engineer preparation on the Warsaw, Praga, approaches pressured Radzievskii to push his army, and operate at night in order to get to Praga before a German build up. The tank army's northwesterly advance threatened to completely encircle German units in the Brest-Litovsk area. The tank army's advance was conducted under increasingly heavy air attacks, and its corps ran low on fuel and ammunition. Radzievskii slowed the advance, allowing fuel and supply trucks to catch up and replenish units.

The tank army's fighting condition in concluding phases of offensive operations were extremely poor. By this time tank armies had suffered already certain losses in manpower, and especially in tanks, the reserves of ammunition and fuel were expended to a considerable degree, and because rear services had fallen behind, these reserves could not be replenished on a regular basis. Coordination, especially with neighboring units and aviation, was disrupted, and sometimes it was lost completely. Tank armies operated often far ahead of the Front's main forces, and reserves were committed already to the engagement and could not always be promptly restored. Conditions of troop control were extremely complex, since tank armies fought on a wide front, and corps carried out different missions, fighting at different depths with large gaps between them devoid of troops. The complexity of the situation in which tank armies were committed against enemy reserves in final phases of operations was also the product of the fact that the countering enemy often enjoyed a superiority in fresh forces.

On 30 July, 3rd Tank Corps became separated from the main forces by 60 kilometers, as a result of which radio communication with it became unstable. The army communications chief pulled a radio relay station out of his reserve, making it possible to achieve continuous communication.[42]

The 2nd Tank Army combat activities in the concluding phase of the Lublin-Brest operation are a instructive example in this respect. Late on 31 July 1944, 2nd Tank Army broke through to Praga but it was unable to capture this industrial center on the move because German troops managed to consolidate their position in the Warsaw fortified area. Radzievskii's tank army, which left the combined-arms armies far behind, exposed its flanks. Its lines of communication stretched hundreds of kilometers, and in a number of places they had been cut by the Germans. Under these conditions, the army went over to defense temporarily, holding it until the approach of the combined arms armies. By 1 August, the Germans brought up five tank and one infantry divisions from different sectors of the front against the 2nd Tank Army and undertook a number of powerful counterstrikes through 5 August, attempting to split the tank army, encircle its parts and destroy them successively. In six days of heavy fighting, the tank army fought all off all German attacks. The tank army's ability to hold made it possible for the left wing of the 1st Belorussian Front to carry out an assault crossing of the Vistula River.

The success of 2nd Tank Army in repelling counterstrikes by a large German panzer force was possible by creating a daring defense. Radzievskii organized the corps in two echelons, creating reserves of massed tank, self-propelled guns and artillery fires on the most important directions. The extensive maneuver of army and corps reserves through a flexible command and control in the course of the engagement allowed a stubborn defense.[43]

On the morning of 29 July, the offensive was resumed after an artillery preparation. As 2nd Tank Army continued to move north, cutting roads and railroads between Warsaw and Bialystok, Radzievskii began receiving reports on the extent of fortifications in the Warsaw suburb of Praga. "To be sure," General Rokossovskii, the Front commander, recalled, "Warsaw was close, we were engaged in heavy fighting on the approaches to Praga. But every step cost a tremendous effort."[44]

Visiting the 2nd Tank Army, General Rokossovskii watched the actions from an observation post on a high factory smoke stack. Peering through a stereoscopic telescope, he could see a panorama of Warsaw and the suburbs. It became apparent to him that to make the final distance was impossible. The character of the German defense was no reflection on Radzievskii's ability to manage the fight. He thought Radzievskii a "capable and energetic general," who repelled the German counterattacks from the Praga area.[45]

"By morning, 31 July," Radzievskii wrote in post war writings, "the tank corps in heavy battle seized a few points around Praga's fortified area. But attempts to penetrate it from the march were not crowned with success. I was forced to give the orders to the army's weakened march units, operating on a front of more than 60 kilometers to transition to defense. Our decision was confirmed by the Front commander."[46]

By evening 31 July, Radzievskii knew his army had run into a significant concentration of German armor units. Believing the considerable strength of German panzer units with their latest heavy "King Tiger" tanks and the tank army at the end of their Lublin-Brest operation, Radzievskii issued orders for the tank units not to attack the fortified area of Praga.

The next day, 2nd Tank Army transitioned to defense. An especially serious situation arose on 3 August when the Germans executed a strong counterstrike against the army's right flank. However, "strong resistance by our tankers and with air support we stood and repelled all their attempts to throw our army from its position," Radzievskii observed.[47]

The tank army fought in cooperation with the 47th Army, which had liberated Siedlce and was pushing German forces to the northwest from the town. For Rokosovskii, "A difficult situation had developed there. The troops of two armies were thinly stretched out along a front facing north, their last reserves committed to action, and the bottom out of the Front reserves as well."[48]General Rokossovskii advanced the 1st Belorussian Front right wing to stabilize the situation.

Suffering heavy losses, 284 tanks and self-propelled assault guns, 40% of which were irretrievable, 2nd Tank Army on 7 August was put in the Front reserve for reconstitution. During the ten days of the Lublin-Brest operation, the tank army advanced 250 kilometers to the outskirts of Warsaw.

General Bogdanov returned from his wound recovery, and Radzievskii resumed his position as Chief of Staff.

M. M. Litvyak, the tank army's political affairs officer, recalled his first meeting with Radzievskii, on the eve of the Vistula-Oder operation. "He was stately in presence, medium height, brunette with intelligent face. His athletic build looked good in a well-fitted and pressed uniform."

Radzievskii organized and fully used his time. He kept himself on schedule, cutting short conversation with Litvyak after dinner. As an evening break during preparatory periods, he used an interrogator for prisoners of war to help him study German.[49]

Radzievskii was noted for his wide erudition, high "Shtabnaya cultura (staff culture)." Before the war he had studied in the Frunze and General Staff Military academies, and his operational staff work was considered by many without equal in the army. In general, he was considered a genuine military "nit-picker." He was highly valued within the Front command. During the Lublin-Brest operation, he commanded well, according to the Army Political officer, Latyshev.

In the middle of October 1944, Radzievskii conducted a review of previous operations with the corps and brigade staffs. He presented a lecture. The political affairs officer who listened to his presentation was carried away by the breadth of his operational-tactical understanding, his finished thoughts and words, and ability to carefully analyze the course of the operation, extracting everything new and instructive that could be useful for subsequent combat operations. Radzievskii spoke easily, and did not make use of analytic notes or summaries. Often he applied to making a staff scheme, showing the concept of developing and executing the operation. Deep in content, his brilliant in report format testified to his uncommon diligence, and his creative treatment of affairs. Such lectures, with thorough analysis of preparations and conduct of combat operations, with practical conclusions and lessons for the chief of staff, were made following each accomplished army operation - Vistula-Oder, East Pomerania, and Berlin.[50]

Remembering such detail in the course of an offensive battle, and because of the great volume of work, Radzievskii clearly had insufficient time, in order to keep an eye on daily news in periodic print. He carved cut out little pauses; sometimes asked the political affairs officer to send to him lectures so that he would be informed. He loved to ask, "what is happening in the wide world?"[51]

On 20 November 1944, the 2nd Tank Army received its Guards designation for actions during the Lublin-Warsaw operations.

In December 1944, Radzievskii attended a conference between representatives from the tank army and pilots from aviation units to explore better organization in cooperation between ground and air fighters. Radzievskii, now General Lieutenant, monitored the working conference which wanted to identify those elements of the cooperation, such as a continuous readiness in communications with different units for the upcoming offensive operation across Poland.

The Vistula-Oder operation preparation began the second half of November 1944. The 1st Belorussian Front staff conducted wargames with subordinate army chiefs of staffs, operations sections, and chiefs of services and branches. Preparing an offensive operation during the war included an entire complex of measures carried out by the commander, staff, political organs, chiefs of branches of arms and services, and rear services. Most important out of these measures was making commander decisions and assigning force missions. Along with this, great attention was paid to organization of coordination and all-round support of the operation.

The importance of tank armies was to achieve surprise and their employment was kept secret during operational planning. Its time schedule, the Front commander allocated tasks to the army commander with an initially small circle of individuals permitted to plan the operation. Usually these were the members of the Military Council and the chief of staff of the army.[52]

Secrecy in transmitting issued instructions was achieved by multiple means, including the most dependable security means, liaison officers traveling to troop units.

In developing the concept of operation and plan, the chief of staff organized the receipt of necessary data from Headquarters branches, chiefs of combat and combat support arms, and combat service support within their functional duties. Conferences for close coordination between the ground units, particularly tank armies and aviation units, were conducted. Coordinating efforts, forces, and equipment involved in an operation occupied a central place in the activity of the army commander and staff and other control organs. Methods and forms of organizing coordination depended upon specific circumstances, as well as the professional training of officials, their combat experience, and other factors. The main method of organizing coordination was the through detailed wargaming of possible situational episodes and corresponding actions of forces. Subordinate large unit commanders, chief of staffs, artillery division commanders, as well as chiefs of arms and services, participated in the wargaming. If time was inadequate, the army commander, after assigning missions, immediately gave instructions on the procedure for troop coordination while accomplishing the most important missions.

Heading the staff, Radzievskii looked at the different variants for action during the course of the upcoming offensive operation. In case of last ditch resistance by German units in Warsaw, he wanted to turn part of the force, 1st Mechanized and 9th Tank Corps, to the east and strike Warsaw from the west. If German forces withdrew, he would use only the 1st Mechanized Corps, and he would have them go on the defensive blocking German actions towards the west and establishing a new defensive line. Which course of action to choose depended upon the German's intentions.[53]

On 10 January, Radzievskii and his staff had completed plans for the operational contingencies, coordinating questions of cooperation with rifle armies and air, reconnoitering march routes, crossings in bridgehead, organizing service regulators, and road guides to crossings, conducting preparation of forces for march from concentration area to the assembly area, and such deception and security of the forces that was necessary for success.

Looking ahead at the area of operation, Radzievskii's staff was busy. A primary consideration, as pointed out by Radzievski, was "the territory between the Vistula and Oder flowed a large number of small rivers and canals, such as Bzura, Pilitsa, and Warta, which crossed their path moving from east to west."[54]

A corps commander recalled, "On the basis of the tank army commander's orders, its instructions on coordination and the plan worked out by the army staff headed by General Lieutenant Radzievskii, and so concluding from the order and instructions corps commanders and staff worked out force crossing tables for crossing the Vistula, planning table for crossing the Pilitsa with two variants, and, also, a table of internal corps radio communications with the 9th and 12th Guards Tank Corps and with the 61st Rifle and 5th Shock Armies."[55]

On 14 January, 1st Belorussian Front assault groups went into action. In two days, lead units drove a wedge 25 kilometers into the German defense. Forward detachments made especially wide use of night actions to develop an offensive, which led to considerable success. In the Vistula-Oder operation, the 9th Guards Tank Corps' forward detachment was committed to the breakthrough at 1700 hours on 15 January. By 2000 hours, it had captured the city of Grojec in a surprise attack during darkness. On the following evening, it cleared Germans out of Zyrardow, and, by 2300 hours, it began the battle for the city of Sochaczew, cutting German westward withdrawal route from Warsaw. Fighting predominately at night, within a single day this forward detachment moved nearly 90 kilometers.

Forward detachments of 9th Guards Tank Corps, and a reinforced 47th Guards Tank Brigade, pursued successfully German forces in the direction of Sochaczew during the operation, advancing 90 kilometers on 16 January. The 12th Guards Tank Corps forward detachment, a reinforced 49th Guards Tank Brigade, advanced 85 kilometers.

The 2nd Guards Tank Army crossed the Pilitsa River, attacking the German's Warsaw grouping. By the next evening, reports came back from the forward units that in the initial battles for the towns of Gruets and Mshchonov, German soldiers used against tankers a new gun, called panzerfaust, a bazooka-type weapon. The antitank gun made holes any place in the tanks and self-propelled assault guns armor. A Red Army inspection team was impressed with the new weapon's simplicity of construction, use, and effectiveness.

General Bogdanov determined that draft combat instructions on how the gun operated and how Germans tactically employed it needed to be sent down to the lower tactical units. A group of Radzievskii's staff officers, including the army chief of technical services and political workers, inspected street battle scenes and interrogated a captured German panzerfaust instructor. Radzievskii coordinated the publication of the combat instructions. The document contained requirements to raise soldier awareness, improve reconnaissance, and quickly notify forces about the appearance of the new gun, and in the course of the battle to immediately take under fire likely panzerfaust positions. The 2nd Guards Tank Army combat instruction gave concrete recommendations for dealing with the new guns, and more importantly, the information was passed to the soldiers without delay.[56]

Throughout the war, the Red Army showed a quick, pragmatic ability to adapt to new weapons systems and tactical techniques as exemplified by Radzievskii's published instructions to counter the panzerfaust. The Red Army established an impressive system using their best officers to collect and study their war experiences for immediate conversion into directives and combat instructions. This is how an army improves and adapts itself under fire.

By the evening of 16 January, Radzievskii's staff learned that the corps' forward detachments had captured Sokhachev, advancing well to the rear of the German's Warsaw grouping. His staff considered different actions for the offensive operation. Once again, if the Germans resisted to the last, he planned to turn the 1st Mechanized and 9th Tank Corps to the east and attack the Germans from the west. If the Germans withdrew, he would keep the 1st Mechanized Corp advancing and cover the army's flank and rear. Which of these various options to choose would depend upon the German's intentions.[57]

On 17 January 1945, Litvyak remembered, "In the tense hours, I went to call on General Lieutenant, A. I. Radzievskii. On his work map was a marked line, where our forces occupy, but it was not a continuous line. Other the other hand, blue arrows representing enemy attempts to breakthrough to the west, were noticeably numerous."[58]

According to the Red Army Staff Field Service Regulation 1942, the chief of staff map should at any moment give a clear picture of the general situation, enable him to report on the latest situation, and arrive at a decision, enabling him to assess the situation as estimated by subordinates.

In a maelstrom of command post activities, the chief of staff must remain an objective, calm throttle to any given situation. Litvyak observed, "Aleksei Ivanovich each minute in calculation. The telephone rang non-stop; the army commander and staff asked about the situation. Entered the combat reports from the army's units. Radzievskii collected, calm, business-like. Neither an unnecessary word nor an unnecessary gesture."[59] Radzievskii was the personification of the Red Army's adoption of the Napoleonic epigram, "There are no trifles in battle."

Like all chiefs of staff with too many tasks and uncooperative subordinates, Radzievskii maintained a reputation as summed by another army commander who described him as always, ". . . his friendly smile, smart and dashing . . ." appearance. "Highly educated, talented military leader, perfectly skilled in his business. In spite of high demands, Aleksei Ivanovich Radzievaskii always remained a peaceful, charming man."[60]

Marshal I. S. Konev, remarked, "The staff, headed by General Radzievskii, made the impression of a very well coordinated organism, in which the commander can fully place his reliance. In the staff's operation was not allowed failures and mistakes." Konev added, "this was a finely organized army, in combat terms preparation, organization, and in operations notably courageous, forcefully and great activeness."[61]

"Somewhere near midnight the assistant chief of staff for the 5th Motorcycle Reconnaissance Regiment reported that it ran into an organized defense with self-propelled guns. Avoiding battle, the regiment found a way through to the southwest. Radzievskii instructed the 1st Mechanized Corps to commit the army reserve, the 9th Heavy Tank Regiment. These instructions were approved by General Bogdanov," Litvyak recalled concerning Radzievskii's decisiveness and support by his commander.[62] The political worker, Litvyak noted Radzievskii was "very active, always down to the units," exhibiting an involvement in operations and running a unit much like a German General Staff officer.

In the staff flowed the information. On one occasion, a report raised an alarm; a German infantry battalion moving around the open right flank of the 6th Heavy Tank Regiment approached the army's command post along the highway, where it located, standing a few kilometers away, a truck convoy with ammunition, fuel, and food. No available units under hand, neither a training, engineer, nor reconnaissance unit, Radzievskii decided to throw into the battle a specially designated, attached wheel amphibian unit, 274th Separate Motorized Battalion.[63] He had managed the defense of his command post several times during the war; he remained remarkably cool. Later that evening, with additional forces, the situation was cleared up.

On 21 January, orders from the Front commander required the seizure of Bromberg. This town, as an objective, fixed the attention of the staff and commander of the tank army. For conducting the deep reconnaissance, Radzievskii sent in its direction the 16th Separate Motorcycle Battalion which operated as the army's reconnaissance detachment.

Initial attempts to capture Bromberg from the south failed. Bogdanov and Radzievski had to redirect the army around the town, attacking it from the west. The town was taken on 23 January. General Bogdanov notified the

Front on the lack of fuel. In a postwar article Radzievskii noted, "in the Vistula-Oder operation 2nd Tank Army for sixteen days of offensive waited for fuel in general some five days (30% of the time)."[64] In spite of the logistical restraints, the 2nd Guards Tank Army's aggressive offensive actions, which lasted sixteen days and developed to a depth of 700 kilometers, were ensured chiefly by the outstanding condition of the material, good preparation of the combat equipment and personnel for the combat activities, and adequate support of the troops in all respects.[65]

As a rule, each army had to prepare, by 2000 hours, a field report to be given to the Front. The chief of staff would put together these reports which gave in brief time, location, condition, composition, disposition of friendly and enemy units, and what the commander was doing or intended to do. These reports were signed by the commander, the commissar, and the chief of staff.[66]

On 1 February, 2nd Guards Tank Army was directed to assist the 2nd Belorussian Front in its East Pomeranian operation. "We received the Front commander's directive in which our army was ordered on that day to turn the front and inflict a counterstrike to the north for the advancing forces, Army Group Vistula, and destroy it in coordination with other front forces," Radzievskii recalled.[67]

To turn an army in a new direction is difficult to carry out easily and quickly. Its forces were conducting combat on a line of 40 kilometers. And, in this situation, great organizing work was performed by the army staff. The staff officers, putting aside for the time maps of the Berlin direction, in a short period prepared a regrouping plan. Almost all the officers of the operations sections were ordered to the corps. Radzievskii departed for the 9th Tank Corps, which was the first to execute the difficult maneuver.

Army staff representation decided different issues. They helped the corps staff quickly bring the combat orders to the brigades, lead out units on new march routes, determined together with corps staff formation changes and taking combat sectors, periods cleared for moving units, etc. At this time in the corps and brigades were located groups of political workers.[68] The 2nd Guards Tank Army participated in the operation which lasted a month before allowing the tank force to pause and refit.

On the eve of the Berlin operation, in the preparation period, was concluded summaries of the winter offensive. With a similar analysis of the force's operations the leadership of the army, corps, brigades, and separate units was conducted by Radzievskii. Similar measures held in all units and army subordinates. In widely popularized positive examples, he understood the most instructive episodes of combat action, and he revealed the reasons for failure in them, in order not to repeat them in the future.

Preparation for the upcoming operation went according to plan. Despite limited organizing time, staffs worked out different tactical issues. A majority of time was spent preparing units to operate in larger populated cities and organizing coordination.

In mid April 1945, on the eve of the Berlin operation, Marshal Zhukov, commander of the 1st Belorussian Front, made last minute checks on the actions of his subordinate commanders. He wanted to ensure close cooperation between forward rifle armies making the penetration of the enemy defense and the commitment of the tank armies through the defense. He called the 1st Tank Army commander who had already been forward to check with the 8th Guards Army in order to finalize coordination. Zhukov phoned the 2nd Guards Tank Army headquarters and asked for Bogdanov. He was not at the headquarters. He, too, had gone to the 5th Shock Army command post, the sector through which his tankers would be committed. Radzievskki took the phone call. Zhukov questioned the whereabouts of the corps commanders who had been designated for action in the assault. "They are up front with Kuznetsov (commander 3rd Shock Army) people in connection with the forthcoming job," Radzievskii replied.[69]

"I could only marvel not at our tank commanders, how they had raised their operational and tactical skills during the war," Zhukov wrote.[70]

By the sixth day of the Berlin operation, 19 April 1945, 2nd Guards Tank Army units penetrated the Oder defensive line and broke through to the northeastern outskirts of Berlin. The 9th Guards Tank Corps bypassing to the north, captured Nauen on 24 April. The following day, the tank corps enveloped the left flank of the Germans' Berlin forces. The same day, lead forces of the 4th Guards Tank Army linked with the 2nd Guards Tank Army, completing the encirclement of Berlin on 25 April 1945.

Postscript

After the war, A. I. Radzievskii commanded a number of higher level units. He became the first deputy chief of the General Staff Military Academy, then the Chief of Main Directorate Training Schools (1959-1969). In 1969, he moved to the post Chief of the Frunze Military Academy. During the period, he wrote highly incisive studies of Red Army operations in such works as *Breakthrough* and *Tank Strike*, directly extending his legacy. Under his general editorship at the Frunze Military Academy, he worked on a major military research studies entitled *Army Operations* and the four volume *Tactics in Combat Example* (Division, Regiment, Battalion, Company).

In 1978, General Radzievskii was moved to a retiring position held by senior officer but assigned work as Inspector of the Inspector General of the Ministry of Defense. He died 30 August 1979, leaving behind a rich legacy of his combat experience and keen analytical studies, which have been a guiding basis in force development and training for the Soviet and, now, Russian Armies.

NOTES:

1. Red Army General Staff, *Sbornik materialov pizucheniyu opyta voiny, No. 2,* September-October 1942, (hereafter *Sbornik No. 2*), Moscow: Voenizdat 1942, p. 86.
2. Ibid. pp. 85-86.
3. Red Army General Staff, *Field Service Regulation,* (Hereafter *FSR '42*) Moscow: Voenizdat, 1942, paragraph 10.
4. K. Rokossovsky, *A Soldier's Duty,* Moscow: Progress Publishers, 1985, p. 60.
5. M. G. Grigorenko, *Skvoz ognennye vikhri,* Moscow: Voenizdat, 1987, p. 71.
6. *Sbornik No. 2,* p. 77.
7. Ibid.
8. Rokossovsky, p. 69.
9. *Sbornik No. 2,* p. 77.
10. D. Z. Muriev, *Proval operatsii "TAIFUN",* Moscow: Voenizdat, 1966, p. 195.
11. Rokossovsky, p. 92.
12. L. M. Sandolov, *Na moskovskom napravlenii,* Moscow: Nauka, 1970, p. 269.
13. A. I. Radzievskii, *Proryv,* Moscow: Voenizdat, 1979, p. 21.
14. Sandolov, p. 273.
15. Radzievskii, p. 21.
16. *Sbornik No. 2,* p. 66.
17. Ibid., p. 64.
18. Ibid.
19. Ibid.
20. John Ellis, *Cavalry: The History of Mounted Warfare,* New York: G. P. Putnam's Sons, 1978, p. 184.
21. D. A. Ivanov, *Fundamentals of Tactical Command and Control,* translated and published under the auspices of the U. S. Air Force, 1977, p. 67.
22. *FSR '42,* paragraphs 18, 19, 21.

23. I. M. Anan'ev, *Tankovye armii v nastuplenii,* Moscow: Voenizdat, 1988, p. 164.
24. O. A. Losik, Stroitel'stvo *i boyevoye primeneniye sovetskikh tankovykh voisk v gody velikoi otechestvennoi voiny,* Moscow: Voenizdat, 1979, pp. 289-290.
25. V. Kasatonov, "General armii A. I. Radzievskii," *Voenno-istoricheskii zhurnal,* 7-1981, (hereafter *Vizh 7-81*), p. 92.
26. *FSR '42,* paragraphs 8-11.
27. P. P. Tovstukha and R. M. Portugal'skii, *Upravlenie voiskami v nastuplenii (po opytu velikoi otechestvennoi voiny),* Moscow: Voenizdat, 1981, p. 52.
28. Anan'ev, p. 224.
29. *FSR '42,* para 79.
30. A. I. Radzievskii, editor, *Armeiskie operatsii,* Moscow: Voenizdat, 1977, p. 11.
31. *FSR '42,* para 79.
32. A. Radzievskii, "Na puti k Varshave," *Voenno-istoricheskii zhurnal,* 10-1971, pp. 70-71.
33. Losik, p. 295.
34. Ibid., p. 296.
35. Ibid.
36. Ibid., p. 163.
37. Ibid., p. 183.
38. Radzievskii, *Vizh* 10-1971, p. 74.
39. Losik, pp. 143-44.
40. Radzievskii, *Vizh* 10-1971, p. 74.
41. Anan'ev, pp. 312-13.
42. Ibid., p. 223.
43. Ibid., p. 371.
44. Rokossovsky, p. 256.
45. Ibid., p. 258.
46. Radzievskii, *Vizh* 10-1971, p. 75.
47. Ibid.
48. Rokossovsky, p. 258.
49. M. M. Litvyak, *Porodnennye bronei,* Moscow: Voenizdat, 1985, p. 11.
50. Ibid., p. 13.
51. Ibid.
52. Losik, p. 287.
53. Litvyak, p. 70.
54. A. Radzievskii, "Stremitel'nye deistviya tankovykh armii," *Voenno-istoricheskii zhurnal,* 1-1965, p. 11.
55. M. D. Solomatin, *Krasnogradtsy,* Moscow: Voenizdat, 1963, p. 138.
56. Litvyak, pp. 68-69.

57. Ibid., p. 70.
58. Ibid., p. 73
59. Ibid.
60. G. G. Semenov, *Nastupaet udarnaya,* Moscow: Voenizdat, 1986, p. 258.
61. I. S. Konev, *Zapiski kommanduyushchego frontom,* Moscow: Voenizdat, 1981, p. 171.
62. Litvyak, pp. 73-4.
63. Ibid., p. 74.
64. A. I. Radzievskii, *Tankovyi udar,* Moscow: Voenizdat, 1977, p. 236.
65. Anan'ev, p. 391.
66. *FSR '42,* para 59.
67. Radzievskii, *Vizh* 1-65, p. 14.
68. Litvyak, p. 183.
69. G. Zhukov, *Reminiscences and Reflections,* Vol 2, Moscow: Progress, 1985, p. 363.
70. Ibid.

INDEX